Synthesis Lectures on Visualization

This series publishes on topics pertaining to scientific visualization, information visualization, and visual analytics. Potential topics include, but are not limited to scientific, information, and medical visualization; visual analytics, applications of visualization and analysis; mathematical foundations of visualization and analytics; interaction, cognition, and perception related to visualization and analytics; data integration, analysis, and visualization; new applications of visualization and analysis; knowledge discovery management and representation; systems, and evaluation; distributed and collaborative visualization and analysis.

Emily Wall

Human Bias in Visual Data Analysis

A Synthesis of Research to Empower Decision Makers

Springer

Emily Wall
Department of Computer Science
Emory University
Atlanta, USA

ISSN 2159-516X ISSN 2159-5178 (electronic)
Synthesis Lectures on Visualization
ISBN 978-3-032-09306-6 ISBN 978-3-032-09307-3 (eBook)
https://doi.org/10.1007/978-3-032-09307-3

This Springer imprint is published by the registered company Springer Nature Switzerland AG
The registered company address is: Gewerbestrasse 11, 6330 Cham, Switzerland

Acknowledgements

This book wouldn't be possible without the input, feedback, and contributions of Mengyu Chen, Yanan Da, Thomas Davidson, Ziwei Dong, Robert Jarman, Shiyao Li, and Keke Wu.

Contents

Introduction

1

1.1 What Is Visual Data Analysis?

The ubiquity and utility of *data* has transformed decision-making in a wide variety of domains: technology companies analyze product usage patterns to inform future releases; medical researchers examine patients' biological markers in clinical trials to understand the effectiveness of a treatment; retailers make decisions about the distribution of products to stock based on past sales trends, and so on. Increasingly, these decisions are aided both by (1) machine learning (ML) models that make predictions from past observations and (2) data visualizations that offload some human cognitive tasks to visual perception. Collectively, these mechanisms inform data-driven decisions in a process we will refer to as **visual data analysis**. Visual data analysis is an end-to-end process comprised of processes for data collection and modeling, visualization of data and model outputs, and ultimately resulting in decision-making outcomes.

Example. Consider the following example, which can serve as context for the remainder of this book. Suppose Catherine is on the undergraduate admissions committee for a large university. The university receives tens of thousands of applicants every year and can only admit a fraction of them as the new cohort. The admissions committee members, including Catherine, employ a visual data analysis process to help in making their admissions decisions and ensure they ultimately make fair and consistent decisions for students with similar qualifications. Their admissions process uses both machine learning and visualizations. Catherine first reviews each individual application, including viewing visualizations of how the applicant compares to the rest of the application pool with respect to test scores, GPA, etc. Then, she consults the model output before finalizing an applicant's admissions decision. The machine learning model was constructed by data scientists and is used to predict admissions decisions of candidates based on historical admissions data. Catherine uses the model like a "sanity check"—in cases of disagreement with her decision and the model, she uses that as a cue to further review the applicant before finalizing her decision.

E. Wall, *Human Bias in Visual Data Analysis*, Synthesis Lectures on Visualization,
https://doi.org/10.1007/978-3-032-09307-3_1

1.2 What Could Go Wrong?

Informed by data with all of the autonomy of a human arbitrating final decisions, what could possibly go wrong in visual data analysis? Combining computational and perception-based human approaches in visual data analysis is often perceived to be optimal in leveraging the strengths of both humans and machines [5]. However, both machines and humans share a common problem: **bias**.

We use the term "bias" loosely to refer to *sources of error or noise in decision-making that can be introduced by both humans and machines*. Biases are present throughout the visual data analysis process—humans collect, clean, and filter data; visualizations and machine learning models make assumptions and simplifications about data; and humans use their judgment and intuition when making decisions with models and visualizations of data. However, it is common at each phase to take the previous steps for granted by assuming that the data, visualization, models, and decision-makers are fair and accurate. We argue that these assumptions demand greater scrutiny when developing systems for visual data analysis, particularly when using visualizations and machine learning together. Because humans and machines are inextricably connected to each phase in this process, there are a number of biases and uncertainties that may arise and propagate through visual data analysis [2, 8]. When perpetuated by both machines and humans, a dangerous cycle can lead to upholding or even intensifying of systemic biases.

For instance, zooming into the machine learning aspects of the process of visual data analysis, we observe a recent surge in AI-generated content. Alongside the rise in AI model use is an especially concerning trend of increasing instances of AI failures [1]. Examples include data scientists building models that perpetuate racial biases [6] or deny critical healthcare coverage [4].

Similarly, if we scrutinize the human and visual components of visual data analysis, humans are subject to a number of biases. For instance, people are subject to perceptual biases that influence the way they view visualizations, such as perceiving the correlation between values in a scatterplot to be stronger than they are when the data are aligned with one's prior knowledge [9]. This is a perceptual artifact of a broader bias known as confirmation bias [7] wherein individuals generally weigh information preferentially when it aligns with their pre-existing beliefs, and down-weigh it when it disagrees with their pre-existing beliefs.

These instances highlight a need to re-evaluate the role of data, humans, machine learning, and visualizations in these processes, including sources of bias and potential methods to mitigate them. It also suggests a potential function of visualizations as a medium through which users may interact with models, with an added layer of awareness. Part 1 describes these concepts in greater detail.

Example. Continuing with the example of Catherine, there could be a number of sources of bias throughout her process of visual data analysis. For instance, the data that the model is trained on may be biased, e.g., in the time period of applications used to train the model.

Data to train a model from 2021 applicants may look quite different than data to train a model from 2025 applicants, due to the impact of the COVID-19 pandemic on admissions processes. Likewise, if not designed with fairness in mind, the model may exhibit biases against applicants according to their race, gender, or ability. When viewing visualizations of the applicants' credentials, Catherine may be subject to perceptual biases that make unfounded trends appear salient, and when making final admissions decisions, Catherine can be "anchored" on the first applicant she reviews, which can influence her review, positively or negatively, of the subsequent applicants just due to the order that she viewed them [3]. These and numerous other biases may infiltrate the process of visual data analysis.

1.3 Where Do We Go from Here?

Work addressing the identification and mitigation of bias in ML and humans has largely taken place in relatively disjoint research communities. Bias and fairness in ML has been primarily addressed by researchers in computing, while bias in humans is typically addressed by researchers in the cognitive and social sciences.

In this book, we assert that as ML systems and visualizations become more intertwined in visual data analysis, understanding the interplay between machine and human biases becomes critical for designing responsible and fair data-driven systems.

In the remainder of this book, we will provide essential background on the types of bias that occur in computational and human systems (Part 1), efforts to characterize or measure those biases (Part 2) as well as methods for mitigating or reducing them (Part 3), and conclude with a look forward to research to come (Part 4). The chapters contained herein provide both high-level background information on a topic (sections prefaced with *Background:*) or take an in-depth dive into a particular research finding (sections prefaced with *Research:*). Throughout this book, we will detail both the disparate efforts related to computational and human biases, as well as more recent efforts that begin to bridge the gaps between the two.

References

1. *AI Incident Database*. URL: https://incidentdatabase.ai.
2. Geoffrey Ellis. *Cognitive biases in visualizations*. Springer.
3. Mussweiler Englich and T Mussweiler. "Anchoring effect". In: *Cognitive Illusions: Intriguing Phenomena in Judgement, Thinking, and Memory* (2016), p. 223.
4. Virginia Eubanks. *Automating inequality: How high-tech tools profile, police, and punish the poor*. St. Martin's Press, 2018.
5. T.M. Green, W. Ribarsky, and B. Fisher. "Visual analytics for complex concepts using a human cognition model". In: *2008 IEEE Symposium on Visual Analytics Science and Technology* (2008), pp. 91–98. DOI: https://doi.org/10.1109/VAST.2008.4677361.

6. Josh Lovejoy and Jess Holbrook. *Human-Centered Machine Learning*. URL: https://medium.com/google-design/human-centered-machine-learning-a770d10562cd.
7. Raymond S Nickerson. "Confirmation bias: A ubiquitous phenomenon in many guises". In: *Review of general psychology* 2.2 (1998), pp. 175–220.
8. Dominik Sacha et al. "The role of uncertainty, awareness, and trust in visual analytics". In: *IEEE transactions on visualization and computer graphics* 22.1 (2016), pp. 240–249.
9. Cindy Xiong, Lisanne van Weelden, and Steven Franconeri. "The curse of knowledge in visual data communication". In: *IEEE transactions on visualization and computer graphics* (2019).

Part I

Define: Computational, Cognitive, and Metacognitive Deficits

2 Process Models in Visual Data Analysis

2.1 Background: Machine Learning Process Models

Process models have been developed in machine learning, identifying the common themes and patterns practitioners follow when developing models. Amershi et al. [1] observed software engineering teams applying ML in practice, and in combination with prior experiences developing data science tools, present a common workflow which we base our discussion on below (Fig. 2.1).

The first step is to formulate the problem one wishes to solve into an ML task by describing the problem space and deciding what types of data and models will be useful [1, 5]. The next step is data curation, including tasks such as data collection, cleaning, labeling, and feature engineering [1, 5, 10]. There is often significant human influence at this stage, e.g., crowdworkers can be used for labeling.

After data has been prepped, the modeling phase begins. Data scientists and machine learning experts use various techniques to decide on appropriate models to fit to the data and optimize for a given evaluation metric, e.g., accuracy, recall, or F1-score [1, 10]. After rounds of model iteration including selecting, training, and evaluating models on testing data subsets, the model is often deployed in the real world where end-users interact with and use its predictions [10]. These downstream users could be the ones that trained the model or new users with no concept of the training data or model limitations.

An alternative model of data science was introduced by Crisan et al. [3] using the analogy of "passing the data baton" between phases (Fig. 2.2). This model includes four high-level stages of Preparation, Analysis, Deployment, and Communication, each of which is further decomposed into constituent steps. For instance, the Deployment stage is comprised of Monitoring and Refinement, which steps back to the Analysis phase, while the Communication stage is comprised of Dissemination and Documentation. The analogy in the Crisan et al. paper [3] makes it natural to consider propagation of biases as inherited problems when one is handed the "baton."

E. Wall, *Human Bias in Visual Data Analysis*, Synthesis Lectures on Visualization,
https://doi.org/10.1007/978-3-032-09307-3_2

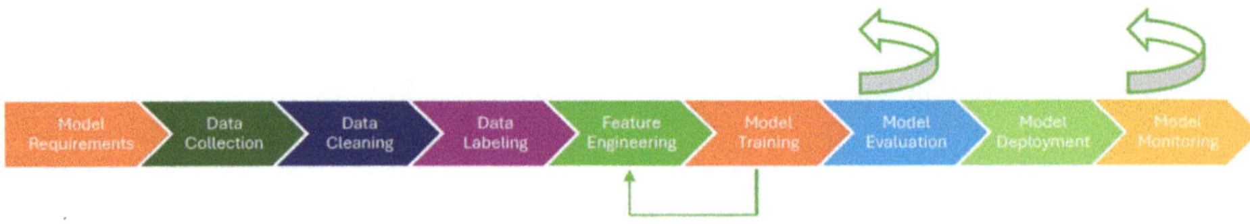

Fig. 2.1 A typical machine learning workflow. Figure based on [1]

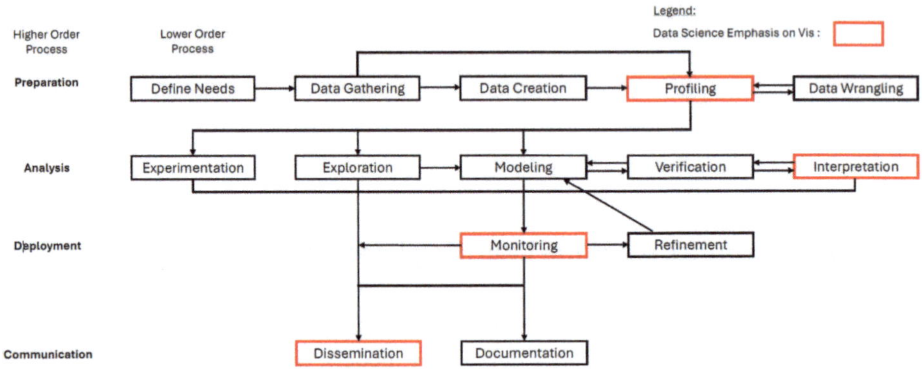

Fig. 2.2 A typical data science workflow. Figure based on [3]

These process models are useful to think about in the context of this book as the steps enumerated in them provide a structured way for us to assess where biases may be introduced in the process and how they may propagate from step to step.

2.2 Background: Cognitive Process Models

Several prevalent theories exist in the decision-making, information processing, and visualization literature for describing aspects of the analytic process [2]. Here we describe a handful of prominent process models. These models provide a glimpse of the mechanisms by which biases may form and how they may influence cognitive processes. For example, the sensemaking process was studied by Pirolli and Card [8]. Sensemaking is a term used in visual analytics to describe the process of learning about data through a visual interface; however, the term more generally refers to the process by which information is gathered, hypotheses are formulated, evidence is extracted, and the hypotheses are evaluated. For visual data analysis, this is a process of exploring the data attributes together with the data model predictions and attempting to explain any patterns against the conceptual models or hypotheses framing the problems of interest.

Pirolli and Card [8] studied this process by performing a cognitive task analysis with intelligence analysts. They proposed that the sensemaking process could be roughly described

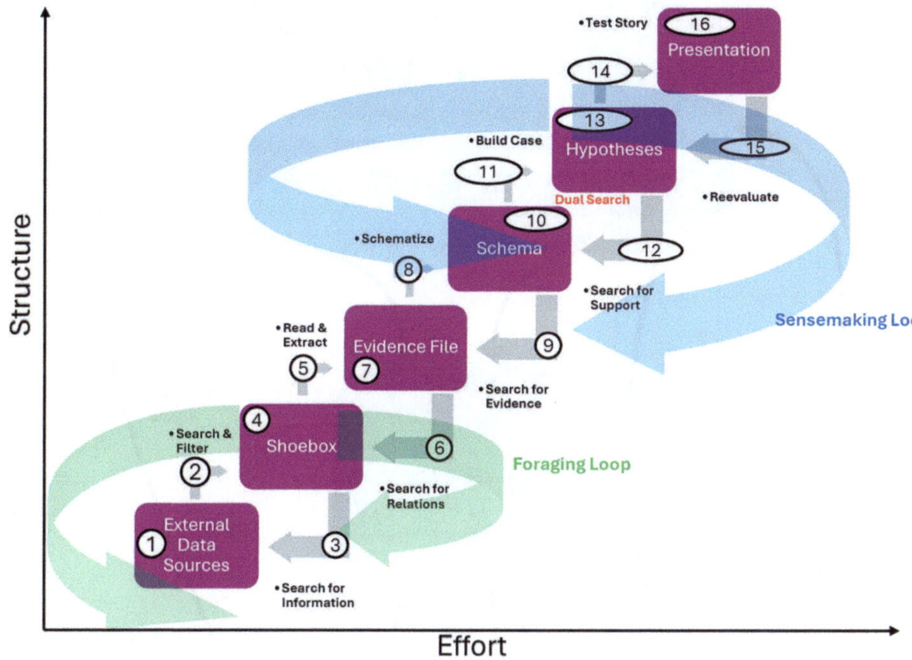

Fig. 2.3 The sensemaking loop. Figure based on [8]

by two loops: (1) a foraging loop to search for information and (2) a sensemaking loop to resolve an understanding of the information (Fig. 2.3). Each of these higher level processes is then decomposed into a series of cognitive actions (e.g., the foraging loop involves iteratively finding evidence from external data sources, compiling the evidence, and then skimming it to look for relevant information).

On the other hand, Klein et al. [7] studied the sensemaking process as an iterative framing and re-framing of information. They postulate that analysts begin with some frame of reference when examining data, then continuously compare, refine, and create new frames throughout analysis to refine their understanding of the data (Fig. 2.4).

Other aspects of the analytic process have also been explored within the context of visual data analysis. Green et al., for example, describe how a human cognition model can be utilized in visual analytics [4]. Their model details the often-complex relationship between human and computer for things like information discovery, creation of new knowledge, and generation and analysis of hypotheses. They describe how tasks and information should be distributed across humans and computers to leverage their complementary strengths in visual analytics (Fig. 2.5).

Similarly, Sacha et al. [9] describe the process of knowledge generation in visual analytics in terms of the related roles of the human and computer. Their model consists of loops for knowledge generation, verification, and exploration (Fig. 2.6).

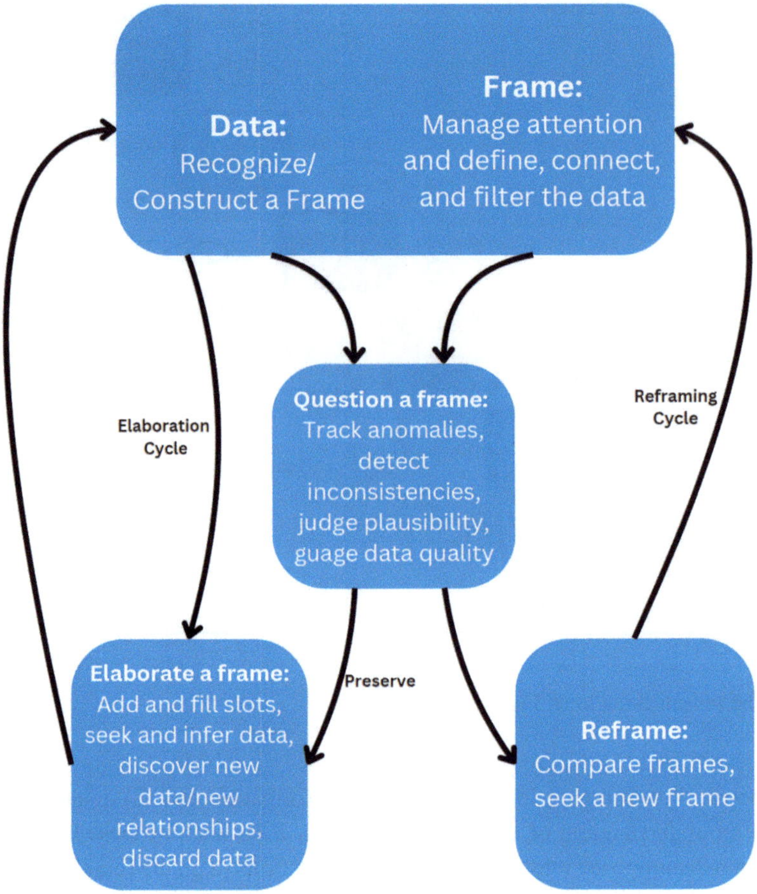

Fig. 2.4 The data-frame model of sensemaking. Figure based on [7]

It is clear from these models that the process of learning and making inferences about data can entail a number of cognitive and perceptual decisions, such as data identification, pattern detection, information discrimination, classification, and selection between discrete options. Multiple types of bias may be introduced into the process by each type of decision, and they may be compounded over the repeated analysis cycles.

2.3 Background: Visual Analytic Process Models

The process models for machine learning (Sect. 2.1) and for cognitive processes (Sect. 2.2) zoom in on the machine and human factors in visual data analysis, respectively. In this section, we detail process models of visualization and visual analytics [2, 4, 6] that focus

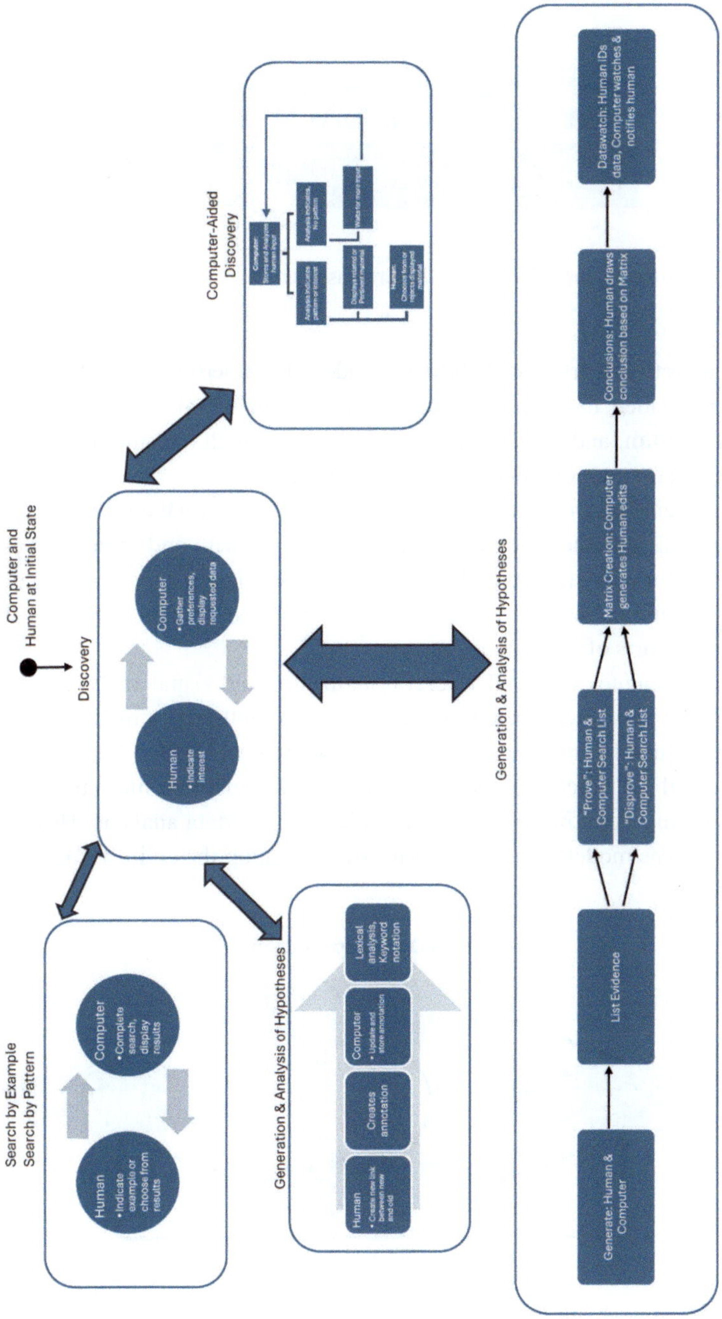

Fig. 2.5 The human cognition model. Figure based on [4]

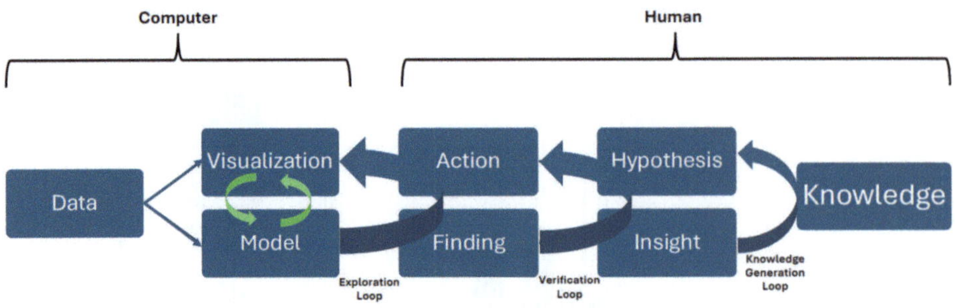

Fig. 2.6 The knowledge generation model. Figure based on [9]

on interactions between artifacts. While the models described in this section all vary in the level of detail provided, they all (sometimes implicitly) include at least *four* key components: **data**, **model**, **human**, and **visualization**, which interact with one another in various ways. We describe the structure of some of these models next.

Humans perceive *visualizations* of *data* and *models*, from which they gain knowledge, draw conclusions, generate new hypotheses, and interact with and refine the visualization. For instance, Keim presents a high-level model of the relationships between *data*, *visualization*, *models*, and *knowledge* [6] (Fig. 2.7), featuring self-loops indicating the iterative nature of several parts of the process.

Booth and colleagues surveyed several visualization and visual analytic process models to-date, including those mentioned thus far, in order to synthesize a more granular model of human decision-making in VA using **dual-process theory** [2] (Fig. 2.8).

Collectively, these process models provide us with structured mechanisms with which to consider and inspect how decisions are made in visual data analysis. However, distinct from these previous models, this book focuses on elevating the entire **visual data analysis**

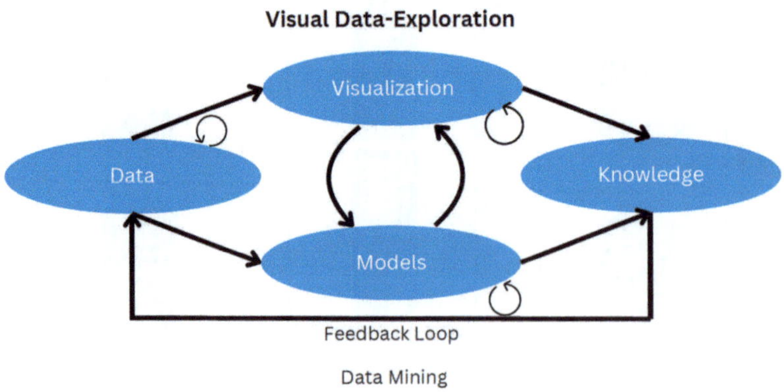

Fig. 2.7 A model of visual analytics. Figure based on [6]

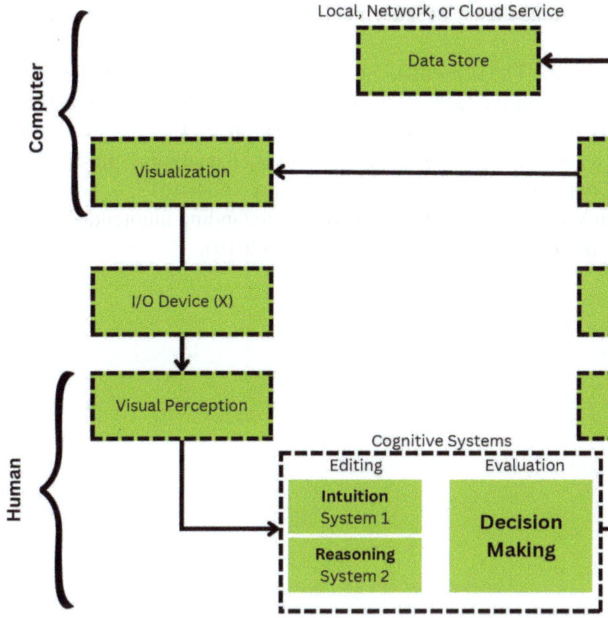

Fig. 2.8 A dual-process model of decision-making with visual analytics. Figure based on [2]

process to understand the nuances and relationships of biases from both *machine* and *human* sources. In the next chapter, we detail sources of bias that may appear throughout these tandem processes in visual data analysis.

References

1. Saleema Amershi et al. "Software engineering for machine learning: A case study". In: *2019 IEEE/ACM 41st International Conference on Software Engineering: Software Engineering in Practice (ICSE-SEIP)*. IEEE. 2019, pp. 291–300.
2. Paul Booth, Nicholas Gibbins, and Spyros Galanis. "Towards a Theory of Analytical Behaviour: A Model of Decision-Making in Visual Analytics". In: *Proceedings of the 52nd Hawaii International Conference on System Sciences*. 2019.
3. Anamaria Crisan, Brittany Fiore-Gartland, and Melanie Tory. "Passing the data baton: A retrospective analysis on data science work and workers". In: *IEEE Transactions on Visualization and Computer Graphics* 27.2 (2020), pp. 1860–1870.
4. T.M. Green, W. Ribarsky, and B. Fisher. "Visual analytics for complex concepts using a human cognition model". In: *2008 IEEE Symposium on Visual Analytics Science and Technology* (2008), pp. 91–98. DOI: https://doi.org/10.1109/VAST.2008.4677361.
5. Charles Hill et al. "Trials and tribulations of developers of intelligent systems: A field study". In: *IEEE Symposium on Visual Languages and Human-Centric Computing*. 2016, pp. 162–170.
6. Daniel Keim et al. *Visual analytics: Definition, process, and challenges*. Springer, 2008. DOI: https://doi.org/10.1007/978-3-540-70956-5_7.

7. Gary Klein et al. "A data–frame theory of sensemaking". In: *Expertise out of context*. Psychology Press, 2007, pp. 118–160.
8. Peter Pirolli and Stuart Card. "The sensemaking process and leverage points for analyst technology as identified through cognitive task analysis". In: *Proceedings of international conference on intelligence analysis*. Vol. 5. McLean, VA, USA. 2005, pp. 2–4.
9. Dominik Sacha et al. "Knowledge generation model for visual analytics". In: *IEEE Transactions on Visualization and Computer Graphics* 20.12 (2014), pp. 1604–1613.
10. Harini Suresh and John V Guttag. "A framework for understanding unintended consequences of machine learning". In: *arXiv preprint* arXiv:1901.10002 (2019).

Defining Biases in Visual Data Analysis 3

3.1 Background: A Pipeline of Biases

In this section, we describe some examples of how biases may occur and propagate throughout the process of visual data analysis, beginning with the data and ending with the human decision-maker, as a primer for the more detailed descriptions of the human-centric biases in the next sections.

Data

Since visual data analysis depends upon initial data, downstream use of data in visualizations or machine learning models to inform decision-making requires careful scrutiny and understanding. Many potential sources of uncertainty or bias may exist in data, for example, measurement precision, completeness (e.g., missing values or aggregated data), credibility, and quality [71].

When data is gathered from the world, it inherently represents the world as it is rather than the world as we may think or want it to be. Suresh et al. refer to this as *historical bias* [75]. For instance, data about hiring decisions is subject to the biases of the people making those decisions [5]. Data representing arrest records are subject to a variety of factors such as racially discriminatory police practices [63]. Assuming no bias in the mechanics of collection, while the data gathered could accurately represent the world, a model or visualization based on biased data will simply replicate and perpetuate these biases into predictions and ultimately surface them to a human decision-maker.

Bias may also be introduced based on how data was sampled. Consider a relatively common practice of convenience sampling [19]. Convenience sampling relies on sampling of data from an easily accessible population that may not be representative of the population of interest. For instance, in human-subject studies, convenience sampling from undergraduate and graduate students at the investigators' university might (incorrectly) be used to

E. Wall, *Human Bias in Visual Data Analysis*, Synthesis Lectures on Visualization,
https://doi.org/10.1007/978-3-032-09307-3_3

draw conclusions about a broader population (e.g., American youth). Suresh et al. call this *representation bias* when the population from which the data was sampled is not the same population that will actually be using the model downstream [75]. In the field of human–computer interaction, for instance, it was found that studies significantly over-represent WEIRD (Western Educated Industrialized Rich Democratic) populations, with 73% of studies from 2016 to 2020 based on Western participant samples which represent just 12% of the world's population.

Bias can be pervasive within data generation due to inherent human and machine involvement [65]. For example, collecting data on citizen voting patterns may be subject to bias due to politically motivated geographic boundaries (e.g., gerrymandering) or producing a dataset about weather patterns may be subject to bias in the algorithmic technique used to sample from large number of measurements. Further, using arbitrary human-provided labels as a proxy for what modelers actually want to predict can result in *measurement bias* [75]. For example, measurement bias may be present when using the number of arrests as a measure of one's likelihood to commit a crime or when using numerical proxies of loan applicant quality to represent an applicant's actual credit-worthiness.

Any and all of these forms of data bias can affect outcomes from visual data analysis. They also may interact with one another. For instance, a convenience sampling of text data from a readily available news outlet may result in representation bias when this data is used to make predictions about the country-wide state of affairs; it may be further exacerbated by measurement bias according to the specific text analysis techniques used as an approximation for civic well-being. Or a binary classification model may still achieve high accuracy in spite of being based on a skewed dataset that does not adequately represent the real world (e.g., 90% of the majority class, combined with an overly simplistic classifier that only provides the majority prediction for all inputs). For an unsuspecting or inexperienced decision-maker who was unaware of how the model was trained, this high accuracy may lead to over-trust in a biased model.

Models

Bias in machine learning is typically defined by a model's output distribution—does a model produce worse outputs for certain demographic groups than others? Examples of biased ML models abound, e.g., in predictive policing [2] and facial recognition [11].

Beyond the data biases we already covered, there are various biases specific to ML arising from both human and machine sources. Because of this fact, it remains inherently difficult to define machine learning bias since bias can originate at any point throughout the ML modeling pipeline, from data collection to deployment. Various metrics exist such as equality of odds [34] and demographic parity [26] that all provide slightly different measures of a "fair" model [7]. It has been shown impossible for a model to fulfill all measures of fairness when different groups in a dataset have different base rates (e.g., positive ground truth label rate in the context of binary classification) [31]. Certain situations may be suited to one fair performance metric over another; however, this choice is often not obvious due to mismatch of human and mathematical notions of fairness [72] and further complicated since these biases often only arise when looking at intersectional subgroups of a dataset [11].

ML models make simplifications and assumptions about the data they are trained on depending on their architectures. During training, *aggregation bias* occurs when a single model is used on a heterogeneous population and is unlikely to fit all subgroups [75]. For example, if a dataset is heavily skewed toward one subgroup then a model can still maintain high overall accuracy while having very poor performance on the minority group. During model evaluation, if the model is compared to benchmarks that do not represent the user population, or performance metrics are not suitable to real-world use, then *evaluation bias* will occur [75]. In such cases, humans may actually *exacerbate* biases and reduce model performance in cases where rapid changes are made without reference to the global context in interactive machine learning systems [92].

In this sense, models may not perform as well as thought during training because real data is far different from the benchmark data that may contain noise and biases. Thus unsuspected errors will occur once the model is deployed with real users.

Visualizations

The primary source of bias in visual representations of data originates from illusions in visual perception and their relationship with specific design choices. Perceptual biases are introduced by the brain's interpretation of visual cues. For instance, we tend to see patterns in the way items are grouped or arranged [45], even when no such pattern exists, that may result in spurious conclusions. People may perceive correlations differently depending on aspect ratio or scale [62] in scatterplots or other visualization types. For instance, a visualization that demonstrates an increasing trend in a line chart that shows the years 2020–2024 spread over 400 horizontal pixels on the x-axis will appear a steeper trend than the same data and time period shown over 800 pixels. Furthermore, the choice of visualization encodings or mark types (e.g., points vs. bars) may influence people's accuracy in data interpretation [18] or the way that we perceive differences in color [76]. People are more accurate at decoding the value of two length or position encodings than they are two area encodings [18].

Certain chart types, particularly those that show data in aggregate forms (e.g., a bar chart vs. a scatterplot), are shown to result in people wrongly assuming that correlations between data attributes are also causal in nature [94]. Individuals are also shown to suffer from the curse of knowledge with visualizations—i.e., that people assume others will perceive data in the same way they do, even if they know others lack the same context [93]. These preconceptions about data can also lead to interpretations of charts consistent with confirmation bias—e.g., someone who believes that environmental regulations have a positive impact on air quality are more likely to interpret a data on the fringe as an outlier and overestimate the strength of correlation in the data that is shown, whereas someone who believes there is no relationship between the variables will perceive the data to have a weaker correlation [96]. When combined with textual data facts, confirmation bias with visualizations has further potential to be exacerbated (if data facts align with prior beliefs) or mitigated (if data facts disconfirm prior beliefs) [49]. This highlights prominent and nuanced reasons why visual communication can fail.

Perceptual biases have also been studied in the context of icon arrays, where data are shown as a series of icons (e.g., filled and unfilled squares arranged in a square grid). Xiong et al. found that the spatial arrangement of filled squares in a grid can impact people's accuracy in perceiving the data [95]. For instance, when asked to estimate the proportion of filled squares in a 10×10 grid, people tend to overestimate the proportion when the filled squares are concentrated in the center of the grid compared to when the filled squares are arranged in rows from the top left or when the filled squares are arranged around the edge of the grid [95]. Likewise, Morais et al. showed that icon arrays that resemble human forms, referred to as anthropographics, can lead to heightened empathy and prosociality [54]. Specifically, they found that representing data on a migrant crisis led individuals to donate more money when the data were represented as anthropographics compared to a standard bar chart representation [54]. Social implications of perceptual biases have also been examined by Holder and Xiong who found that representations that show data in aggregate (e.g., bar charts or dot plots), thus hiding variability, can lead to heightened agreement with personal attributions of data differences compared to visualizations that emphasize variability (e.g., jitter plots or prediction intervals) [36]. For instance, when shown data on wage disparities for waiters and waitresses, people who saw bar charts were more likely to agree with explanations that the difference in wages was attributable to differences in work ethic compared to differences in how expensive the restaurants were [36]. This highlights that perceptual biases prominent with certain types of visualizations can exacerbate harmful stereotyping.

Humans

While we have described a number of ways that human bias can be injected into visual data analysis (including choices involved in the curation and cleaning of data, choices in ML models and parameterizations, and design choices for the visual representation of data), this section describes the ways that end-user decision-making can also be subject to bias. In addition to perceptual biases that influence the way we interpret visual information cues, we also use cognitive heuristics to make faster decisions, which may occasionally lead to errors or bias [81]. Previous work demonstrated that people are significantly more likely to choose a treatment option framed by its success rate compared to its failure rate (framing effects) [82]. For example, if the same treatment is claimed to be successful 60% of the time vs. failing 40% of the time people will choose the treatment framed by its success, even if they are the exact same treatment. Similarly, when asked to judge the probability of an event (e.g., that a cab involved in a hit-and-run was blue in color), people often neglect to account for the base rate (e.g., the probability of confusing the color in dark or rainy weather) and thus produce systematically flawed judgments [4]. Beyond cognitive and perceptual biases, a third class of biases influencing human decision-making result from individuals' cultural and social influences. Racial bias, for example, has been shown to impact hiring decisions. When responding to job ads, resumes with historically White names (e.g., Emily or Greg) received 50% more calls back than resumes with historically Black names (e.g., Lakisha or Jamal), despite containing equivalent credentials [6]. These biases shape the way that we process information to make decisions and hence play an important role in visual data analysis.

Considering Propagation

In visual data analysis, the combination of both humans and machines in VIS + ML systems can introduce any combination of the aforementioned biases. For example, since people are often involved in virtually every stage of model building, the resulting models may be subject to experts' biases or blind spots [61] regarding the appropriate labeling of data, selection of model parameters, and so on. Further, algorithmic and visual perceptual biases can be compounded in some cases. In some dimensionality reduction approaches, cluster sizes and distances between clusters may not mean anything but people still perceive it as so, which could mislead users to make spurious inferences about their data [91].

The biases that propagate through the visual data analysis process can have unintended downstream effects. For example, once an ML model is deployed, it is at risk for *deployment bias* when users incorrectly interpret a model's output [75]. These errors may be rooted in cognition of the human decision-maker if they refuse to believe a model output that is contradictory to a previous belief (confirmation bias) or refuse to see that bias has occurred since they have invested time and resources into building the model (sunken cost fallacy). Additional cognitive issues arise when people's perception of ML in general mars their ability to reason with model output objectively (mis-calibrated trust). This can cause someone to over-trust or completely disregard the model output.

Example. Consider the earlier example of Catherine from Chap. 1. Suppose Catherine collects high school students' resumes with the goal of improving Computer Science department undergraduate admissions processes. However, due to prior historical biases in department admissions, there is an imbalance in the amount of data for traditionally underrepresented groups including women, racial minorities, and students with disabilities. Subsequent use of this data leads to propagation of such biases into algorithmic decision-making aids. This can be further exacerbated if inappropriate visualizations are used (e.g., axes that do not start at 0 may amplify perceptual differences in grade-point average (GPA) between students who are similarly qualified) or visualizations of the data are used that may mis-align with admissions goals (e.g., visualizations that may focus only on numerical measures of applicants may systematically suppress certain types of students). The committee may elect to train an ML model to predict the USA college GPA (on a scale from 0 to 4) of applicants as an indicator of how competitive they are, based on prior GPA data collected from admitted students. Yet if this model choice is overly simplistic, it may further ignore the differences between groups in the data and have poor accuracy for underrepresented subgroups. Furthermore, by electing to use GPA as the target variable, the committee is using a model that is predicting something that is inherently different from their original objective. Admitting quality students goes beyond simply having a good college GPA, but it is easier to quantify GPA than perhaps a student's overall quality. When department admissions committees use these tools to ease the manual burden of evaluating applications, the ultimate decisions may be further subject to the individuals' implicit biases related to gender, race, etc., or algorithmic output may reinforce or confirm more explicit biases ingrained in committee members. Collectively, these biases influence the way we perceive and process information at all phases of visual data analysis. By the time information reaches the human decision-maker, it is often unclear what biases may have been injected into the data, visualization, or algorithms that the decision-maker relies on.

This section has given a summary of biases in the visual data analysis process. The next sections delve into more granular definitions and examples of *human biases*, specifically cognitive and metacognitive deficits.

3.2 Research: (Three) Perspectives on Human Bias in Visual Data Analysis

To reach the ultimate goal of developing methods to detect and mitigate bias in visual data analysis, we first must understand what is meant by the term "bias." Originally published at the Workshop on Dealing with Cognitive Biases in Visualizations [88] at IEEE VIS in 2017 and later published as a book chapter in Cognitive Biases in Visualizations [89], this section describes three different ways that the term "bias" is used in reference to different human concepts in visual data analysis. Note that in the original publication, there were (four) perspectives—the fourth being "bias as a model mechanism" used to characterize the other three forms of bias. Please see the workshop paper or book chapter for details on this perspective.

Cognitive, behavioral, and social sciences have described many ways bias can occur in people's analytic processes [44, 59, 67], decision-making strategies [10, 23], and other behaviors. Motivated by the overloaded use of the term "bias" to describe different models and concepts, we describe three different ways people tend to think about or refer to human bias that are relevant in the context of visual data analysis. These perspectives include (1) bias as a cognitive processing error, (2) bias as a filter for information, and (3) bias as a preconception. These three perspectives are not mutually exclusive; rather, they present different, potentially overlapping perspectives on bias relevant in the context of visual data analysis.

Example

Consider the following example. Suppose Susan is using a visual data analysis tool to explore possibilities for purchasing a new home. She uses the tool to browse photos, explore different areas of the city, and refine her understanding of what features of a home are important to her. From her exploration, she intends to go view the homes in person and ultimately make a purchasing decision. Throughout the following sections, we will describe how each perspective on bias can impact Susan's process and visual data analysis in general. For each of the three perspectives, we provide a brief description, present an example scenario, and discuss how these perspectives inform and influence visual data analysis.

3.2.1 Bias as a Cognitive Processing Error

Description

From heuristics and bias research, bias is an error resulting from an unconscious deviation from rational behavior. Cognition is frequently conceptualized as a dual process [13]. The two processes are often termed "intuition" and "reason" [42], the former being responsible for making quick, automatic decisions, and the latter being responsible for making deliberate, reflective decisions. It is one's quick judgments that are subject to these types of errors.

Stanovich and West referred to the two cognitive processes as system 1 (intuition) and system 2 (reason) [73]. In this analogy, system 1 is largely subconscious and prone to making errors (bias), while system 2 is responsible for recognizing and correcting errors through intentional deliberation. These types of errors result from shortcuts in cognition, broadly referred to as heuristics [42]. Bias then is described as the method or mechanism by which the error occurs. However, the process of heuristic decision-making does not always lead to errors; it usually facilitates fast decision-making.

Example

From this perspective, there are dozens of types of bias. One such example is anchoring bias [81], which refers to the tendency to be heavily reliant on an initial value or anchor. It is analogous to a center of mass: people are unlikely to strongly deviate from their center. In Susan's home-buying scenario, she will likely be subject to anchoring bias during the price negotiation of her purchase. That is, the home's initial list price forms an anchor point and will thus subconsciously impact the amount she is willing to offer. Susan's offer for the home might have been very different had she made an offer given a different initial list price. She might even pay more money for the same home due to the tendency not to strongly deviate from the anchor point. Systems apprised of probable cognitive errors like anchoring bias have the potential to help users make better decisions by guarding against such errors, providing appropriate counterexamples, or by suggesting other ranges of data values that a user might consider.

Relevance to Visual Data Analysis

Common heuristic errors include confirmation bias [57] which describes the way people tend to accept confirmatory evidence of a pre-existing hypothesis and dismiss contrary information. Another common error is availability bias [80], where people tend to rely more heavily on information that is easily remembered (e.g., most recent). Similarly, the attraction effect [37] describes the tendency for a decision to be influenced by an inferior alternative. Collectively, these errors shape the way people search for and interpret information and hence have implications for visual data analysis. In fact, Dimara and colleagues [22] categorized 154 different types of these biases into 7 categories relevant to visualization, including errors related to causal attribution, recall, and estimation, among others, while Boonprakong et al. [8] describe how cognitive biases are studied more broadly in the HCI community. Several papers have also studied specific cognitive biases. For instance, Dimara et al. [21] showed that the attraction effect is present in users of information visualizations, while others have shown that priming and anchoring effects can be replicated in visualizations and visual data analysis [17, 84, 87]. Later in this book, we will discuss the case of confirmation bias as well [49]. Hence, bias impacts users outside of laboratory decision-making studies and can lead to incorrect decisions and inefficiencies in visual data analysis.

3.2.2 Bias as a Filter for Information

Description

Bias acts as a filter through which we manage and perceive information. The challenge of information overload [52] motivates this analogy. Information overload, now commonly leveraged in consumer research to influence purchasing behavior, refers to a point beyond people's cognitive and perceptual limits where performance and decision-making suffer [51]. Under overload conditions, people selectively allocate attention and other mental resources to the tasks or information of highest priority. One's filter or bias thus determines what information is gathered and how sensory information is distinguished and interpreted [32].

The literature on goal-directed attention and resource allocation posits that all perception is guided by top-down influences, such as the allocation of endogenous attention [27, 60, 77]. Top-down perception governs which sensory information is identified in a scene based on goals. Bias does not make for a purely objective filter for information, however. Heuer refers to perception as an "active" process that "constructs" reality [35]; this is in contrast to a passive process that simply records reality. Similarly, obvious or important information is sometimes filtered out. For example, in one classic selective perception task, participants were shown video footage of people wearing either white or black shirts passing a basketball. Participants were asked to count how many times white-shirt basketball players on a team passed the ball to each other [70]. Most participants count the appropriate number of passes but about half fail to perceive a glaringly misfit player walk across the court. The misfit player is in a black outfit, and is consequently treated as part of the task that is selectively ignored while attention is focused on the white-shirt players. In contrast to top-down perception, bottom-up perception refers to the way external factors influence attention [64]. When there is a loud noise or someone says your name across the room, you notice despite top-down attentional and perceptual focus. Visual attention can be similarly grabbed by flashing, movement, or other visual cuing in a display.

Example

In our home-buying scenario, Susan may experience information overload [52] as she explores homes on the market in a visual data analysis tool. She might see hundreds of homes available in the area, each with dozens of attributes. Thus, her filter or bias will govern which information she perceives and which she dismisses. For example, she may only select to view single-family homes, removing condominiums, town homes, and apartments from the visualization. If removed, some options that may be relevant to Susan's other search criteria will not be visibly available, though still in the underlying data and system. An intelligent system aware of Susan's attentional capacity may want to make some of that information known at an appropriate point in the analytic process. By leveraging knowledge about people's perceptual strengths and limitations, mixed-initiative systems could present information in ways that are easy for users to understand and at a time when mental resources are available.

Relevance to Visual Data Analysis

A great deal of research in visual perception [83] has been leveraged by researchers in information visualization and visual data analysis to present information in ways that are most perceptually accessible [29]. Preattentive processing theory [79], for example, describes the nature and limits of visual information processing. In creating visual representations of data, this is often used by designers as a guide to prevent overwhelming a user's perceptual limitations. Similarly, Gestalt principles [45] refer to the relationships inferred by the visual system based on proximity, groupings, symmetry, etc. between visual elements. Thus, understanding how people's filters work can inform things like which visual widgets or elements to place in close proximity to one another or which graph layout algorithm is most appropriate. Indeed, Patterson et al. [58] listed supporting attention and user mental models as some of the key visualization leverage points for design grounded in human cognition. Other studies have shown the accuracy of decoding information when it is shown to users in visual [18] or tactile formats [97], supporting effective visualization and physicalization [41] design. Recently, Strain and colleagues demonstrated that some perceptual biases in scatterplots can be mitigated by manipulating the size and opacity of points around trend lines [74].

3.2.3 Bias as a Preconception

Description

Analysts approach mixed-initiative systems bringing all their experiences and internal influences that unconsciously shape their approaches to the analysis process. This, in turn, influences the ways they interact with systems. The consequence is that the user model within the system, the analytic products, and provenance may be shaped by each individual's unconscious biases. These types of bias may seem to have little to do directly with the task at hand. Yet, because they shape the person, there is a high likelihood they can influence mixed-initiative sensemaking.

Unconscious biases arise in a number of ways. They derive from a person's cultural beliefs and traditions, which include their implicit assumptions and expectations regarding stereotypes [33] and can affect the way people make decisions. Like the example mentioned in the previous chapter, a 2004 study revealed that implicit racial bias led to 50% more calls back for job interviews for equivalent resumes with traditionally white versus black names (Emily/Greg vs. Lakisha/Jamal) [6]. Unconscious biases can also result from general self-confidence or self-esteem, as well as comfort or familiarity level with the capabilities of a machine's analytics and interface functions. Related personality traits render some people more risk seeking or risk averse, shaping how they push boundaries exploring a space of hypotheses or push the capabilities of the computational system. These characteristics are thus seen as a source of individual variability between people.

Example

Susan is avoiding listings for houses downtown in the city. Having lived in the suburbs for many years, Susan assumes that neighborhoods near downtown have higher crime rates and lower economic stability. She believes she should not make a housing investment there. The availability of recent census results and police reports within the real estate analytic tools enable Susan to explore her assumptions and refine her thinking. A mixed-initiative system may detect her avoidance of downtown properties and could prompt her to challenge her assumptions with the related data.

Relevance to Visual Data Analysis

Unconscious biases shape analysts' assumptions and stereotypes about analytical tools and mixed-initiative aids, and they shape assumptions and stereotypes about the data/analytical subjects (e.g., presumed reliability or trustworthiness of certain sources). Implicit attitudes shape the formulation of hypotheses and the questions about the assumptions and the consequences of those hypotheses. Klein and colleagues posited that the entire sensemaking process begins with a practitioner framing the problem, and the selected framework, however minimal, then shapes what an analyst thinks about and what structure they think with [44].

Frames reflect a perspective an analyst takes to make sense of data or to solve a problem. As implicit attitudes shape an analyst's perspective, they shape the analyst's frames, thereby shaping the sensemaking process. In one recent example, my students and I demonstrated that implicit gender biases influence the way reviewers rate the competitiveness of university applicants, but also showed that a simple intervention using selective visual highlighting of certain text (e.g., adjectives that were more commonly used to describe men vs. women [24]) mitigated the disparities caused by the bias [20].

Because they are built to record a number of different types of user behaviors throughout the analysis process, mixed-initiative systems may be particularly well positioned to aid in the assessment of unconscious biases of analysts. We argue that it is possible for a mixed-initiative system to capture and integrate unconscious, preconception biases into analytics through the user model and track those biases through user interactions and changes in the mental model over time. In fact, that is precisely what my collaborators did in our paper, where we captured, characterized, and mitigated implicit biases related to the gender, gender, and party affiliation of politicians in a 2021 user study [90] (described further in Sect. 7.5).

3.3 Background: Metacognitive Deficits

In addition to the three types of human bias (bias a cognitive processing error, bias as a filter for information, and bias as a preconception), there is another category of *metacognitive deficits*, only recently considered in VDA. The term metacognition was coined by John Flavell in his seminal work to refer to *"knowledge and cognition about cognitive phenomenon"* [30]. The majority of metacognition research occurs in educational psychology, with experimental focus developing learning strategies in topics such as mathematics [99], reading comprehension [28], and programming [50]. Some interventions leverage *disfluency*, where perceptual expectations are disrupted using, e.g., a non-traditional font [47], which leads to use of more conscientious thinking [1]. Interventions in education help individuals develop better strategies for learning; likewise, we posit that metacognitive interventions in visual data analysis [16] can help analysts learn or reinforce skills on-the-job, rather than relying on separate training courses.

Metacognitive deficits have been studied in a broad range of contexts. For instance, the Dunning–Kruger effect is a well-studied phenomenon wherein highly skilled individuals have a systematic tendency to *underestimate* their abilities, while their lower skilled counterparts systematically *overestimate* their abilities [46]. This phenomenon has been shown in contexts including mathematics, English proficiency, and even humor [46]. Metacognitive deficits are also a notable clinical characteristic of autism spectrum disorder [53], particularly in the context of social and emotional awareness, and are at the root of a number of learning disabilities [78].

In a broader sense, "meta" has been applied to other terms including metamemory [56] or metareasoning [66] to mean a concept that is self-referential (i.e., memory of memory or reasoning about reasoning, respectively). Metacognition then refers to *"thinking about thinking"*; that is, a suite of skills related to **constituent skills** of *knowledge of cognition* and *regulation of cognition* [68]. *Knowledge of cognition* can be divided into at least three types of knowledge: *declarative* (knowledge "about"), *procedural* (knowledge of "how"), and *conditional* (knowledge of "when" and "why") [9]. *Regulation of cognition* is the enactment of *knowledge of cognition* to exercise control of one's thinking [9]; it relies on three essential skills [3, 40, 69]: *planning* (selecting strategies), *monitoring* (awareness of comprehension and task performance), and *evaluating* (assessing the outcomes of learning). Successful metacognitive interventions can bolster one or more of these skills.

For instance, Schraw [68] describes ways to improve *knowledge of cognition*, including strategies such as *skimming*, *slowing down*, and *activating prior knowledge*. These strategies represent *declarative* knowledge, how to use them represents *procedural* knowledge, and when and why to use them represents *conditional* knowledge. On the other hand, improving *regulation of cognition* can involve the use of regulatory checklists. King found that fifth graders who employed such checklists outperformed others across a variety of problem-solving measures [43]. These successful uses of metacognitive training have analogs in visual data analysis, described more in Chap. 7.

In visualization research, at least four notable research thrusts are relevant to metacognition, although seldom through the theoretical lens of metacognition. First, research on learning analytic tools often involves testing dashboard visualizations to help individuals self-monitor progress (e.g., [14, 25, 48, 85]) or prompt reflection (e.g., [50]). Second, visualizing uncertainty (e.g., [39]) is often motivated by helping individuals adjust their process and confidence levels in data, a form of self-monitoring [98]. However, studies have shown that while uncertainty visualizations may aid judgments, they do not improve metacognitive efficiency [12]. Third, perceptual disfluency and other forms of "visual difficulties" may be beneficial to engage analytical reasoning when interacting with visualizations [38], although these techniques have yet to be empirically evaluated. Finally, in prior work, my colleagues and I have introduced techniques to promote real-time *awareness* of potential biases in visual data analysis processes [55, 86, 90] and, most recently, characterized a specific metacognitive deficit in VDA, Dunning–Kruger effect [15], detailed further in Sect. 5.6.

References

1. Adam L Alter et al. "Overcoming intuition: metacognitive difficulty activates analytic reasoning." In: *Journal of experimental psychology: General* 136.4 (2007), p. 569.
2. Julia Angwin et al. "Machine Bias". In: *ProPublica* (2016).
3. Alice F Artz and Eleanor Armour-Thomas. "Development of a cognitive-metacognitive framework for protocol analysis of mathematical problem solving in small groups". In: *Cognition and instruction* 9.2 (1992), pp. 137–175.
4. Maya Bar-Hillel. *The base-rate fallacy in probability judgments*. Tech. rep. DECISIONS and DESIGNS INC MCLEAN VA, 1977.

5. Solon Barocas and Andrew D Selbst. "Big Data's Disparate Impact". In: *Cal. L. Rev.* 104 (2016), p. 671.
6. Marianne Bertrand and Sendhil Mullainathan. "Are Emily and Greg more employable than Lakisha and Jamal? A field experiment on labor market discrimination". In: *American economic review* 94.4 (2004), pp. 991–1013.
7. Alex Beutel et al. "Putting Fairness Principles into Practice: Challenges, Metrics, and Improvements". In: *AAAI/ACM Conference on Artificial Intelligence, Ethics, and Society (AIES)*. 2019.
8. Nattapat Boonprakong et al. "How Do HCI Researchers Study Cognitive Biases? A Scoping Review". In: *Proceedings of the 2025 CHI Conference on Human Factors in Computing Systems*. 2025, pp. 1–20.
9. Ann Brown. "Metacognition, executive control, self-regulation, and other more mysterious mechanisms". In: *Metacognition, motivation, and understanding* (1987).
10. Eli T Brown et al. "Finding waldo: Learning about users from their interactions". In: *IEEE Transactions on visualization and computer graphics* 20.12 (2014), pp. 1663–1672. DOI: https://doi.org/10.1109/TVCG.2014.2346575.
11. Joy Buolamwini and Timnit Gebru. "Gender shades: Intersectional accuracy disparities in commercial gender classification". In: *Conference on Fairness, Accountability and Transparency*. ACM, 2018, pp. 77–91.
12. Olivia Burton et al. "Visualization of uncertainty aids spatial judgments but fails to improve metacognitive efficiency". In: *Proceedings of the Human Factors and Ergonomics Society Annual Meeting*. Vol. 61. 1. SAGE Publications Sage CA: Los Angeles, CA. 2017, pp. 1390–1393.
13. Shelly Chaiken and Yaacov Trope. *Dual-Process Theories in Social Psychology*. New York: Guilford Press, 1999. ISBN: 9781572304215.
14. Li Chen et al. "Design of Learning Analytics Dashboard Supporting Metacognition." In: *International Association for Development of the Information Society* (2019).
15. Mengyu Chen, Yijun Liu, and Emily Wall. "Unmasking Dunning-Kruger Effect in Visual Reasoning and Visual Data Analysis". In: *IEEE Transactions on Visualization and Computer Graphics* (2024). to appear.
16. Mengyu Chen et al. "A Novel Lens on Metacognition in Visualization". In: (2025).
17. Isaac Cho et al. "The Anchoring Effect in Decision-Making with Visual Analytics". In: *IEEE Conference on Visual Analytics Science and Technology (VAST)* (2017).
18. William S Cleveland and Robert McGill. "Graphical perception: Theory, experimentation, and application to the development of graphical methods". In: *Journal of the American statistical association* 79.387 (1984), pp. 531–554.
19. William G Cochran. *Sampling techniques*. John Wiley & Sons, 2007.
20. Yanan Da et al. "Visual Salience to Mitigate Gender Bias in Recommendation Letters". In: *Workshop on Visualization for Social Good (VIS4Good, at VIS'23)* (2023).
21. Evanthia Dimara, Anastasia Bezerianos, and Pierre Dragicevic. "The attraction effect in information visualization". In: *IEEE transactions on visualization and computer graphics* 23.1 (2017), pp. 471–480.
22. Evanthia Dimara et al. "A task-based taxonomy of cognitive biases for information visualization". In: *IEEE transactions on visualization and computer graphics* (2018).
23. Wenwen Dou et al. "Recovering Reasoning Process From User Interactions". In: *IEEE Computer Graphics & Applications* May/June.March (2009), pp. 52–61. URL: http://citeseerx.ist.psu.edu/viewdoc/download?doi=10.1.1.157.407%7B%5C&%7Drep=rep1%7B%5C&%7Dtype=pdf.
24. Kuheli Dutt et al. "Gender differences in recommendation letters for postdoctoral fellowships in geoscience". In: *Nature Geoscience* 9.11 (2016), pp. 805–808.
25. Erik Duval. "Attention please! Learning analytics for visualization and recommendation". In: *Proceedings of the 1st international conference on learning analytics and knowledge*. 2011, pp. 9–17.

26. Cynthia Dwork et al. "Fairness through awareness". In: *Innovations in Theoretical Computer Science Conference*. 2012, pp. 214–226.
27. Howard E Egeth and Steven Yantis. "Visual attention: Control, representation, and time course". In: *Annual Review of Psychology* 48.1 (1997), pp. 269–297.
28. María Rosa Elosúa et al. "Improving reading comprehension: From metacognitive intervention on strategies to the intervention on working memory executive processes". In: *Universitas Psychologica* 12.SPE5 (2013), pp. 1425–1438.
29. Jean-Daniel Fekete et al. "The value of information visualization". In: *Information Visualization* (2008), pp. 1–18.
30. John H Flavell. "Metacognition and cognitive monitoring: A new area of cognitive–developmental inquiry." In: *American psychologist* 34.10 (1979), p. 906.
31. Sorelle A Friedler, Carlos Scheidegger, and Suresh Venkatasubramanian. "On the (im) possibility of fairness". In: arXiv:1609.07236 (2016).
32. John P Frisby and James V Stone. *Seeing: The Computational Approach to Biological Vision*. Cambridge, MA: The MIT Press, 2010.
33. Anthony G Greenwald and Linda Hamilton Krieger. "Implicit bias: Scientific foundations". In: *California Law Review* 94.4 (2006), pp. 945–967.
34. Moritz Hardt, Eric Price, Nati Srebro, et al. "Equality of opportunity in supervised learning". In: *Advances in Neural Information Processing Systems*. 2016, pp. 3315–3323.
35. Richards J. Heuer Jr. *Psychology of Intelligence Analysis*. Washington, DC, 1999.
36. Eli Holder and Cindy Xiong. "Dispersion vs Disparity: Hiding Variability Can Encourage Stereotyping When Visualizing Social Outcomes". In: *IEEE Transactions on Visualization and Computer Graphics* 29.1 (2022), pp. 624–634.
37. Joel Huber, John W Payne, and Christopher Puto. "Adding asymmetrically dominated alternatives: Violations of regularity and the similarity hypothesis". In: *Journal of Consumer Research* 9.1 (1982), pp. 90–98.
38. Jessica Hullman, Eytan Adar, and Priti Shah. "Benefitting infovis with visual difficulties". In: *IEEE Transactions on Visualization and Computer Graphics* 17.12 (2011), pp. 2213–2222.
39. Jessica Hullman et al. "In pursuit of error: A survey of uncertainty visualization evaluation". In: *IEEE transactions on visualization and computer graphics* 25.1 (2018), pp. 903–913.
40. Janis E Jacobs and Scott G Paris. "Children's metacognition about reading: Issues in definition, measurement, and instruction". In: *Educational psychologist* 22.3-4 (1987), pp. 255–278.
41. Yvonne Jansen et al. "Opportunities and challenges for data physicalization". In: *proceedings of the 33rd annual acm conference on human factors in computing systems*. 2015, pp. 3227–3236.
42. Daniel Kahneman and Shane Frederick. "A model of heuristic judgment". In: *The Cambridge Handbook of Thinking and Reasoning* (2005), pp. 267–294.
43. Alison King. "Effects of training in strategic questioning on children's problem-solving performance." In: *Journal of Educational psychology* 83.3 (1991), p. 307.
44. Gary Klein et al. "A data–frame theory of sensemaking". In: *Expertise out of context*. Psychology Press, 2007, pp. 118–160.
45. Kurt Koffka. *Principles of Gestalt Psychology*. Vol. 44. Routledge, 2013.
46. Justin Kruger and David Dunning. "Unskilled and unaware of it: how difficulties in recognizing one's own incompetence lead to inflated self-assessments." In: *Journal of personality and social psychology* 77.6 (1999), p. 1121.
47. Tim Kühl and Alexander Eitel. "Effects of disfluency on cognitive and metacognitive processes and outcomes". In: *Metacognition and Learning* 11.1 (2016), pp. 1–13.
48. Derick Leony et al. "GLASS: a learning analytics visualization tool". In: *Proceedings of the 2nd international conference on learning analytics and knowledge*. 2012, pp. 162–163.

49. Shiyao Li et al. "Confirmation Bias: The Double-Edged Sword of Data Facts in Visual Data Communication". In: *Proceedings of the 2025 CHI Conference on Human Factors in Computing Systems*. ACM, 2025.
50. Dastyni Loksa et al. "Programming, problem solving, and self-awareness: Effects of explicit guidance". In: *Proceedings of the 2016 CHI conference on human factors in computing systems*. 2016, pp. 1449–1461.
51. Naresh K Malhotra. "Information load and consumer decision making". In: *Journal of Consumer Research* 8.4 (1982), pp. 419–430.
52. James T Milord and Raymond P Perry. "A Methodological Study of Overloadx". In: *The Journal of General Psychology* 97.1 (1977), pp. 131–137.
53. Eleni Mitsea, Athanasios Drigas, and Charalampos Skianis. "Metacognition in Autism Spectrum Disorder: Digital Technologies in Metacognitive Skills Training". In: *Technium Soc. Sci. J.* 31 (2022), p. 153.
54. Luiz Morais et al. "Can anthropographics promote prosociality? a review and large-sample study". In: *Proceedings of the 2021 CHI Conference on Human Factors in Computing Systems*. 2021, pp. 1–18.
55. Arpit Narechania et al. "Lumos: Increasing awareness of analytic behavior during visual data analysis". In: *arXiv preprint* arXiv:2108.02909 (2021).
56. Thomas O Nelson. "Metamemory: A theoretical framework and new findings". In: *Psychology of learning and motivation*. Vol. 26. Elsevier, 1990, pp. 125–173.
57. Raymond S Nickerson. "Confirmation bias: A ubiquitous phenomenon in many guises". In: *Review of general psychology* 2.2 (1998), pp. 175–220.
58. R. E. Patterson et al. "A human cognition framework for information visualization". In: *Computers & Graphics* 42 (2014), pp. 42–58.
59. Peter Pirolli and Stuart Card. "The sensemaking process and leverage points for analyst technology as identified through cognitive task analysis". In: *Proceedings of international conference on intelligence analysis*. Vol. 5. McLean, VA, USA. 2005, pp. 2–4.
60. Michael I Posner. "Orienting of attention". In: *Quarterly Journal of Experimental Psychology* 32.1 (1980), pp. 3–25.
61. Emily Pronin, Daniel Y Lin, and Lee Ross. "The bias blind spot: Perceptions of bias in self versus others". In: *Personality and Social Psychology Bulletin* 28.3 (2002), pp. 369–381.
62. Ronald A Rensink and Gideon Baldridge. "The perception of correlation in scatterplots". In: *Computer Graphics Forum*. Vol. 29. 3. 2010, pp. 1203–1210.
63. Rashida Richardson, Jason Schultz, and Kate Crawford. "Dirty data, bad predictions: How civil rights violations impact police data, predictive policing systems, and justice". In: *New York University Law Review Online* (2019).
64. Maximilian Riesenhuber and Tomaso Poggio. "Hierarchical models of object recognition in cortex". In: *Nature Neuroscience* 2.11 (1999), pp. 1019–1025.
65. Yuji Roh, Geon Heo, and Steven Euijong Whang. "A survey on data collection for machine learning: a big data-ai integration perspective". In: *IEEE Transactions on Knowledge and Data Engineering* (2019).
66. Stuart Russell and Eric Wefald. "Principles of metareasoning". In: *Artificial intelligence* 49.1-3 (1991), pp. 361–395.
67. Dominik Sacha et al. "Knowledge generation model for visual analytics". In: *IEEE Transactions on Visualization and Computer Graphics* 20.12 (2014), pp. 1604–1613.
68. Gregory Schraw. "Promoting general metacognitive awareness". In: *Instructional science* 26.1 (1998), pp. 113–125.
69. Gregory Schraw and David Moshman. "Metacognitive theories". In: *Educational psychology review* 7.4 (1995), pp. 351–371.

70. Daniel J Simons and Christopher F Chabris. "Gorillas in our midst: Sustained inattentional blindness for dynamic events". In: *Perception* 28.9 (1999), pp. 1059–1074.
71. Meredith Skeels et al. "Revealing uncertainty for information visualization". In: *Information Visualization* 9.1 (2010), pp. 70–81.
72. Megha Srivastava, Hoda Heidari, and Andreas Krause. "Mathematical notions vs. human perception of fairness: A descriptive approach to fairness for machine learning". In: *ACM SIGKDD International Conference on Knowledge Discovery & Data Mining*. 2019, pp. 2459–2468.
73. Keith E Stanovich and Richard F West. "Advancing the rationality debate". In: *Behavioral and Brain Sciences* 23.5 (2000), pp. 701–717.
74. Gabriel Strain et al. "Effects of Alternative Scatterplot Designs on Belief". In: *Proceedings of the 2025 CHI Conference on Human Factors in Computing Systems*. 2025, pp. 1–12.
75. Harini Suresh and John V Guttag. "A framework for understanding unintended consequences of machine learning". In: *arXiv preprint* arXiv:1901.10002 (2019).
76. Danielle Albers Szafir. "Modeling color difference for visualization design". In: *IEEE TVCG* 24.1 (2017), pp. 392–401.
77. Antonio Torralba et al. "Contextual guidance of eye movements and attention in real-world scenes: The role of global features in object search". In: *Psychological Review* 113.4 (2006), pp. 766–786.
78. Guy Trainin and H Lee Swanson. "Cognition, metacognition, and achievement of college students with learning disabilities". In: *Learning Disability Quarterly* 28.4 (2005), pp. 261–272.
79. Anne Treisman. "Preattentive processing in vision". In: *Computer Vision, Graphics, and Image Processing* 31.2 (1985), pp. 156–177.
80. Amos Tversky and Daniel Kahneman. "Availability: A heuristic for judging frequency and probability". In: *Cognitive Psychology* 5.2 (1973), pp. 207–232.
81. Amos Tversky and Daniel Kahneman. "Judgment under uncertainty: Heuristics and biases". In: *Science* 185 (1974), pp. 1124–1131.
82. Amos Tversky and Daniel Kahneman. "The framing of decisions and the psychology of choice". In: *Science* 211 (1981), pp. 453–458.
83. Shimon Ullman. "Visual routines". In: *Readings in computer vision*. Elsevier, 1987, pp. 298–328.
84. André Calero Valdez, Martina Ziefle, and Michael Sedlmair. "Priming and anchoring effects in visualization". In: *IEEE Transactions on Visualization and Computer Graphics* 24.1 (2018), pp. 584–594.
85. Camilo Vieira, Paul Parsons, and Vetria Byrd. "Visual learning analytics of educational data: A systematic literature review and research agenda". In: *Computers & Education* 122 (2018), pp. 119–135.
86. Emily Wall, John Stasko, and Alex Endert. "Toward a design space for mitigating cognitive bias in vis". In: *2019 IEEE Visualization Conference (VIS)*. IEEE. 2019, pp. 111–115.
87. Emily Wall et al. "A formative study of interactive bias metrics in visual analytics using anchoring bias". In: *IFIP Conference on Human-Computer Interaction*. Springer. 2019, pp. 555–575.
88. Emily Wall et al. "Four Perspectives on Human Bias in Visual Analytics". In: *DECISIVe: Workshop on Dealing with Cognitive Biases in Visualizations* (2017).
89. Emily Wall et al. "Four perspectives on human bias in visual analytics". In: *Cognitive biases in visualizations*. Springer, 2018, pp. 29–42.
90. Emily Wall et al. "Left, Right, and Gender: Exploring Interaction Traces to Mitigate Human Biases". In: *arXiv preprint* arXiv:2108.03536 (2021).
91. Martin Wattenberg, Fernanda Viégas, and Ian Johnson. "How to Use t-SNE Effectively". In: *Distill* (2016). DOI: https://doi.org/10.23915/distill.00002. URL: http://distill.pub/2016/misread-tsne.

92. Tongshuang Wu, Daniel S Weld, and Jeffrey Heer. "Local decision pitfalls in interactive machine learning: An investigation into feature selection in sentiment analysis". In: *ACM Transactions on Computer-Human Interaction (TOCHI)* 26.4 (2019), pp. 1–27.
93. Cindy Xiong, Lisanne van Weelden, and Steven Franconeri. "The curse of knowledge in visual data communication". In: *IEEE transactions on visualization and computer graphics* (2019).
94. Cindy Xiong et al. "Illusion of causality in visualized data". In: *IEEE transactions on visualization and computer graphics* 26.1 (2019), pp. 853–862.
95. Cindy Xiong et al. "Investigating perceptual biases in icon arrays". In: *Proceedings of the 2022 CHI Conference on Human Factors in Computing Systems*. 2022, pp. 1–12.
96. Cindy Xiong et al. "Seeing what you believe or believing what you see? belief biases correlation estimation". In: *IEEE Transactions on Visualization and Computer Graphics* 29.1 (2022), pp. 493–503.
97. Zhongzheng Xu, Kristin Williams, and Emily Wall. "Let's Get Vysical: Perceptual Accuracy in Visual & Tactile Encodings". In: *2023 IEEE Visualization and Visual Analytics (VIS)*. IEEE. 2023, pp. 16–20.
98. Nick Yeung and Christopher Summerfield. "Metacognition in human decision-making: confidence and error monitoring". In: *Philosophical Transactions of the Royal Society B: Biological Sciences* 367.1594 (2012), pp. 1310–1321.
99. Rosetta Zan. "A metacognitive intervention in mathematics at university level". In: *International Journal of Mathematical Education in Science and Technology* 31.1 (2000), pp. 143–150.

Part II

Measure: Computational, Cognitive, and Metacognitive Deficits

Measuring Biases in Computational Systems 4

4.1 Background: Fairness Metrics

ML researchers have devised a number of approaches to combat bias by transforming biased data. The AI Fairness 360 toolkit details approaches to help debias data such as reweighing the training dataset to ensure undersampled populations receive more attention or using optimized data preprocessing to learn fair representations which obfuscate protected attributes (e.g., sex, race) while maintaining a dataset's utility [1]. Such solutions primarily address representation bias as they transform datasets to more equitably cover different groups of individuals. Other approaches have utilized techniques from fields such as differential privacy to help ensure more fair data collection [8]. Historical bias can be encoded in textual data when word embeddings reflect outdated gender stereotypes like the association of "man" with "computer programmer" and "woman" with "homemaker." Techniques that shift the embedding space to be more gender neutral can help combat these historical biases [3].

Beyond these solutions for data biases, there are various other biases within ML arising from both human and machine sources. Because of this fact, it remains inherently difficult to define machine learning bias since bias can originate at any point throughout the ML modeling pipeline, from data collection to deployment. Various metrics exist such as equality of odds [7] and demographic parity [5] that all provide slightly different measures of a "fair" model [2].

It has been shown impossible for a model to fulfill all measures of fairness when different groups in a dataset have different base rates (e.g., positive ground truth label rate in the context of binary classification) [6]. Certain situations may be suited to one fair performance metric over another, however this choice is often not obvious due to mismatch of human and mathematical notions of fairness [9] and further complicated since these biases often only arise when looking at intersectional subgroups of a dataset [4]. Which solutions are most appropriate, however, still requires human input on social context [10].

E. Wall, *Human Bias in Visual Data Analysis*, Synthesis Lectures on Visualization,
https://doi.org/10.1007/978-3-032-09307-3_4

References

1. Rachel Bellamy et al. "AI Fairness 360: An Extensible Toolkit for Detecting and Mitigating Algorithmic Bias". In: *IBM Journal of Research and Development* PP (Sept. 2019), pp. 1–1. DOI: https://doi.org/10.1147/JRD.2019.2942287.
2. Alex Beutel et al. "Putting Fairness Principles into Practice: Challenges, Metrics, and Improvements". In: *AAAI/ACM Conference on Artificial Intelligence, Ethics, and Society (AIES)*. 2019.
3. Tolga Bolukbasi et al. "Man is to computer programmer as woman is to homemaker? debiasing word embeddings". In: *Advances in Neural Information Processing Systems*. 2016, pp. 4349–4357.
4. Joy Buolamwini and Timnit Gebru. "Gender shades: Intersectional accuracy disparities in commercial gender classification". In: *Conference on Fairness, Accountability and Transparency*. ACM, 2018, pp. 77–91.
5. Cynthia Dwork et al. "Fairness through awareness". In: *Innovations in Theoretical Computer Science Conference*. 2012, pp. 214–226.
6. Sorelle A Friedler, Carlos Scheidegger, and Suresh Venkatasubramanian. "On the (im) possibility of fairness". In: arXiv:1609.07236 (2016).
7. Moritz Hardt, Eric Price, Nati Srebro, et al. "Equality of opportunity in supervised learning". In: *Advances in Neural Information Processing Systems*. 2016, pp. 3315–3323.
8. Seth Neel and Aaron Roth. "Mitigating bias in adaptive data gathering via differential privacy". In: arXiv:1806.02329 (2018).
9. Megha Srivastava, Hoda Heidari, and Andreas Krause. "Mathematical notions vs. human perception of fairness: A descriptive approach to fairness for machine learning". In: *ACM SIGKDD International Conference on Knowledge Discovery & Data Mining*. 2019, pp. 2459–2468.
10. Mohammad Yaghini, Hoda Heidari, and Andreas Krause. "A Human-in-the-loop Framework to Construct Context-dependent Mathematical Formulations of Fairness". In: arXiv:1911.03020 (2019).

Measuring Biases in Humans 5

5.1 Background: Psychometrics

Psychometrics is the term used to describe the study of psychological measures [58]. Psychological measures can be used to characterize numerous cognitive and perceptual phenomenon including intelligence [57], personality [86], and mental disorders such as PTSD [11], among others. However, following acknowledgement of the complexity of psychological phenomenon and biases inherent to the process of externalization, each of these measures is met with critique and ongoing debate of their validity.

Some of the most common applications of psychometrics lie within the domains of measuring individuals' intelligence and personality traits. Perhaps one of the earliest published intelligence measures originated in the early twentieth century, known as the Binet–Simon scale [9], consisting of approximately 30 tests over the course of 40 minutes. Tests included measures of language skills, memory, and reasoning, among others. The scale was validated by observing increasing scores as individuals aged and by comparing control subjects against those with known cognitive impairments [9]. Later adaptations of intelligence tests resulted from observations of unfair correlations with gender, race, and culture. For instance, the popular Kaufman Brief Intelligence Test (KBIT) [75], comprised of verbal and non-verbal scales, is commonly used to screen for gifted students considering cultural and gender biases.

In personality testing, one of the most well-known examples is the Myers–Briggs Type Indicator (MBTI) [96]. MBTI relies on self-reflection on a series of questions that categorize the individual according to four dimensions: (i) introversion or extraversion, (ii) sensing or intuition, (iii) thinking or feeling, and (iv) judging or perceiving. However, MBTI is widely criticized in the academic community for its poor validity [13]. A more recent alternative is the Big Five personality inventory which characterizes individuals according to (i) openness to experience, (ii) conscientiousness, (iii) extraversion, (iv) agreeableness, and (v) neuroticism [107]. Numerous comparative meta-analyses have found higher validity and reliability in the Big Five than the MBTI [55].

E. Wall, *Human Bias in Visual Data Analysis*, Synthesis Lectures on Visualization,
https://doi.org/10.1007/978-3-032-09307-3_5

With respect to implicit biases, a popular instrument is the Implicit Association Test (IAT) [65]. The IAT has been applied across various domains including gender [4], race [65], and sexual orientation [15] by measuring reaction time to word and/or image pairs. For example, Greenwald et al. conducted a study on racial biases and found many participants were quicker to associate positive words with White names than with Black names, indicating implicit racial bias [65]. At its core, the IAT characterizes implicit biases by measuring the association that people hold between attributes and concepts. The test asks users to quickly and accurately categorize words or images and measures reaction time, such that faster responses indicate stronger associations than slower responses, suggesting how implicit attitudes can influence cognitive processes and behaviors. In the case of the racial implicit association test, it asks individuals to press one button when presented with the image of an African-American face and another button when presented with the image of a European face. Likewise, the test will ask individuals to press one button when presented with a "good" word such as delight, attractive, or fabulous, and another button when presented with a "bad" word such as disgust, sadness, or hate. Implicit biases are then characterized based on the differences in response times when asked to align a common response to "good" with European and "bad" with African-American versus "good" with African-American and "bad" with European.

IAT has been employed to illuminate implicit attitudes in numerous settings. For instance, Capers et al. [21] measured implicit racial preference in medical school admissions committees using the black–white implicit association test [65] on 140 members of the admissions committee composed of faculty and students. The results showed that all groups (men, women, students, faculty) display significant levels of implicit bias in the form of white preference. Education and training on implicit biases are suggested by the authors to reduce such biases. However, since these biases are unconscious, informing individuals about the existence of implicit biases has apparently limited impact during the decision-making process [37]. In the next section, we discuss alternative measures of biases in the context of visual data analysis, and in Part III we introduce ways to leverage those measures to mitigate observed biases.

5.2 Research: Interaction Metrics

User interaction is central to the effectiveness of visual analytic systems [43, 106, 149]. It is the mechanism by which people and systems communicate about data, allowing people to become an integral part of the data exploration process and steer computational models [17, 84, 136]. Analysts leverage their domain expertise and reasoning skills to explore the data via the user interface. They communicate their intents and questions to the system, realized as guiding analytic models or changing the visualization's parameters.

In this chapter, we argue that user interactions play a powerful second role in addition to shaping analytical models: interactions form an externalized record of users' thought processes. Interactions mark the paths of exploratory data analysis, providing an opportunity to glean insight into a person's reasoning and decision-making processes [99, 109]. Enabling human input via user interaction in human-in-the-loop (HIL) approaches enables insights in many domains, especially where uncertainty is high and human reasoning is a valuable addition to data-intensive computation [44]. However, incorporating human reasoning and analysis into computational models may have unwanted side effects, as detailed in Chap. 1 and Part I of this book. Specifically, human biases can materialize in these user interactions.

The effect of biases in VA systems can have far-reaching effects, influencing the evidence upon which analysts rely and the hypotheses they form. Further, when user interaction in visual analytic tools is intended to guide analytic models, biases might be propagated to and amplified by the underlying computational models. The resulting biased analytic models may ultimately prompt analysts to make incorrect or inferior decisions, or simply echo the users' biases back to them. This constitutes an emergent bias in computational systems [54]. We note that human biases are not all bad nor does use of heuristics always produce errors in reasoning; on the contrary, use of heuristics is often positive, producing quicker and more effective decision-making. Such efficiencies may be useful within an HIL system. Thus, we seek ways to understand how bias arises in HIL analytics, to harness positive effects when useful and mitigate negative effects when they might be damaging.

We hypothesize that when data analysis is supported by visual analytic tools, **analysts' biases influence their data exploration in ways that are measurable through their interactions with the data**. This presents an opportunity to leverage user interactions to detect and assess mental pitfalls in real time during the analysis process. While models exist that incorporate measures of human bias, they rely on the final *products* of cognition (e.g., a final choice decision). This does not allow for the real-time measurement of bias in the decision-making process. Instead, we propose that bias can be detected earlier in an analysis *process*, using metrics applied to the user's interactions. Real-time assessment of cognitive performance can be leveraged for adaptive interfaces, responsive to individualized user needs [10, 51]. However, the critical first step in developing systems for cognitive augmentation or mitigation is construction of conceptual frameworks for detecting and assessing a user's cognitive state [45, 59]. This provides the theoretical basis for interpreting behaviors to provide the right machine-based bias interventions.

In this section, we present theoretical foundations for quantifying indicators of human biases in interactive visual analytic systems and propose six preliminary metrics. These metrics are based on the notions of *coverage* and *distribution*, targeting assessment of the process by which users sample the data space. We propose a way to quantify interactions and a naïve baseline model for an unbiased analysis against which the metrics can be interpreted. We emphasize that our proposed metrics do not map one-to-one onto any particular biases. Rather, they describe behavioral indicators that might result from any number of underlying cognitive biases. The work detailed in this section was originally published at IEEE VIS in 2017 [137].

5.2.1 Formalizing Bias in Visual Data Analysis

In this section, we outline the ways cognitive bias may manifest in the analytic process and discuss relationships between bias indicators and the proposed metrics.

Behavioral Indicators of Bias in Interaction Cognitive bias is a consequence of heuristic decision-making processes that allow people to simplify complex problems and make more efficient judgments [74, 127]. A heuristic is a "rule of thumb" for making an inference, or a strategic way in which information is ignored to get to a decision faster [60]. Heuristics frequently ignore or subconsciously weight certain types of information. As a subconscious cognitive process, heuristics also play an integral role in visual analytics. Concerted efforts have been made to delineate the cognitive biases to which analysts may be susceptible [70]. This provides a starting point for understanding biases in the inference and sensemaking process.

There are dozens of cognitive biases captured in the heuristics and biases literature [60, 73]. The cognitive biases relevant to a set of interactions are dependent on the nature of the task people are performing. We focus herein on the cognitive biases that typically make the evaluation of evidence an effective process. We refer to the evaluation of evidence as the process by which data are determined to be relevant to the analysis process at hand. Heuer [70] describes five types of cognitive biases particularly relevant for evaluating evidence, defined in Table 5.1: *vividness criterion, absence of evidence, oversensitivity to consistency, coping with evidence of uncertain accuracy*, and *persistence of impressions based on discredited evidence* (also known as the *continued influence effect*). Each type of bias, including those in Table 5.1, impacts people's behavior in predictable ways. The third column in the table gives an example of how each given type of bias might specifically influence a user's interactions. For each of these examples, we can compute on several measurable patterns of user interaction, which we refer to as **behavioral indicators of bias** or just **indicators of bias**.

We emphasize that our approach is based on the claim that there is *not* a one-to-one mapping between cognitive biases and the proposed metrics. When a user is biased, we expect to find these patterns in their interactions; however, detecting a particular indicator does not necessarily tell us which type of cognitive bias may have caused the behavioral response. We have diagrammed this relationship between the types of cognitive biases discussed in this work and the set of proposed metrics for measuring indicators of bias in Fig. 5.1. The block arrow between biases and indicators represents a many-to-many mapping, the particulars of which we defer to future work. Here we focus on developing metrics that relate to individual indicators of bias.

We also note that other types of biases outside of the ones listed in Table 5.1 and Fig. 5.1 could also be captured by some combination of interaction metrics, including common biases such as confirmation bias or framing effect. We demonstrate this in Chap. 5.3 for the anchoring effect, where an initial value can create an "anchor point," commonly described in the context of price negotiations. Nevertheless, we chose this framework as it provides a

Table 5.1 Cognitive biases relevant to intelligence analysis [70] that produce the measurable behavioral indicators we focus on in this work

Bias	Description	Interaction manifestation
Vividness criterion	Humans rely more heavily on information that is specific or personal than information that is abstract or lacking in detail	e.g., analyst frequently returns to/interacts with data points that are rich in detail
Absence of evidence	Humans tend to focus their attention on the information that is present, ignoring other significant pieces of evidence that may be missing	e.g., analyst filters out a subset of data, forgets about it, and makes future decisions without accounting for the missing data
Oversensitivity to consistency	Humans tend to choose hypotheses that encompass the largest subset of evidence	e.g., analyst interacts almost exclusively with data that supports the largest encompassing hypothesis, dismissing other data
Coping with evidence of uncertain accuracy	Humans tend to choose to accept or reject a piece of evidence wholly and seldom account for the probability of its accuracy	e.g., analyst filters out data that supports a seemingly unlikely hypothesis, thus fully rejecting it
Persistence of impressions based on discredited evidence	Humans tend to continue to believe information even after it has been discredited (also known as the *continued influence effect*)	e.g., analyst continues to interact with data supporting a hypothesis that has been disproved

reasonable breadth yet well-defined scope of biases with which to demonstrate the use of interaction metrics.

What Can We Measure?

To identify ways in which we might measure bias from interaction data, we need to develop two key pieces of theory: (1) what can be measured and (2) a method of interpreting the measurements.

To address (1), we must identify the sets of possible things that can be measured, from which we can derive metrics. Herein we focus on dyads of {types of interaction} with {objects of interaction}. That is, types of interaction include things like clicks, hovers, and drags afforded by a system that can be explicitly captured by event listeners. Semantically similar interactions supported by other device modalities can be mapped to our proposed metrics, but ultimately need to be bound to event handlers. For our preliminary metrics, objects of interaction currently include data points, attributes, and attribute weights; however,

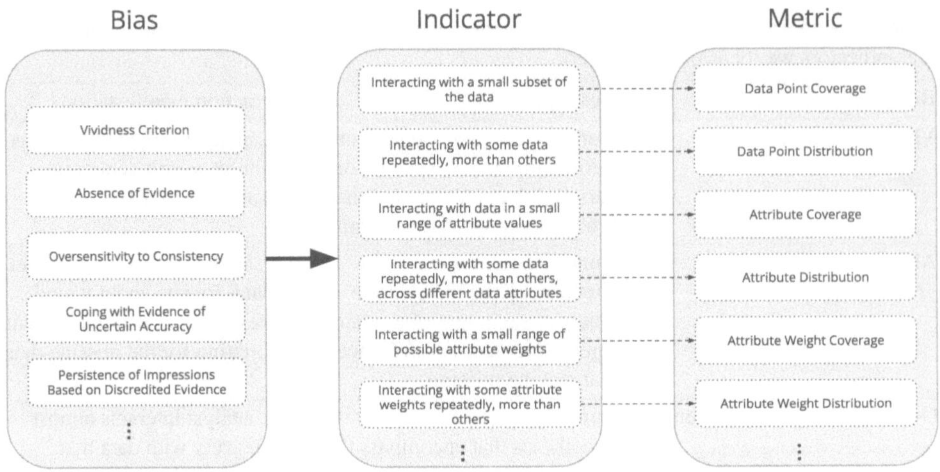

Fig. 5.1 Cognitive biases result in behavioral indicators that are measurable by the proposed metrics. We scope this work to those indicators and metrics depicted above, but there are numerous other biases, behavioral indicators, and ways to measure those indicators

we could conceivably measure interactions with many other objects, including analytic model parameters or interactions with particular views in a multi-view interface. Further, the metrics can only account for the dataset loaded in the system. For example, if an analyst is examining a dataset of criminal suspects, the metrics would not be able to infer about a bias toward a person not represented in the dataset.

To address (2), we must develop baseline models of behavior that would reflect performance under assumptions of non-biased information gathering or decision-making to make appropriate inferences about biased behaviors. We assert that we can formulate models of interaction behavior by conceptualizing the set of data points and possible interactions with those points as a state space over which we can define Markov chains. That is, we let each interaction with a data point be a state in a state space. A user performing that {point, interaction} combination has transitioned to the associated state in the Markov chain. The transition probabilities are the likelihood of subsequent interaction options given the current state or current interaction. For example, if clicking on a point means you are likely to next click on a point in close proximity, the transition probability would be high between those two states. As we will develop further, the dataset defines the points, the interface defines the possible interactions on those points, and together, the visual analytic system defines the state space. Our Markov chain provides a generalizable approach to describing any sequence of interactions with an analytic system. The model can be changed to capture different analytic behaviors by simply altering the transition matrix for the Markov chain on that state space. In this way, we can study different patterns of biased and unbiased behaviors to define relevant baselines for different domains all within a common theoretical framework. But in this work, we will use a simple Markov chain, defined later, making minimal assumptions about what constitutes unbiased behaviors.

Table 5.2 Notation used to describe the bias metrics

Notation	Description
b_μ	Bias metric from the set of all metrics μ, with range $b_\mu \in [0, 1]$, where higher values indicate more prominent indicators of bias
$D = \{d_1, ..., d_N\}$	dataset of size N
$A = \{a_1, \ldots, a_M\}$	set of M attributes describing dataset D
$T = \{click, hover, ...\}$	set of interaction types
D_U	unique set of data points interacted with by the user, where $D_U \subseteq D$
$I(d_n)$	set of interactions with data point $d_n \in D$
$\kappa(X)$	cardinality of set X
$\hat{\kappa}(X)$	expected cardinality of set X, based on a Markov chain model of user interactions
$w = [w(a_1), \ldots, w(a_M)]$	attribute weight vector

To formalize our preliminary metrics, we first define some common notation, which is summarized in Table 5.2. We define $D = \{d_1, \ldots, d_N\}$ to be a dataset of size N. Each data point d_i has a set of M attributes, $A = \{a_1, \ldots, a_M\}$. We define D_U to be the unique set of data points interacted with by a user. $I(D)$ is the set of interactions by a user on the dataset, and $T = \{\text{click, hover}, \ldots\}$ is the set of interaction types. Within a visual analytic system, the set of possible interaction events is $T \cup D$, the union of the set of interaction types afforded by the interface and the set of data points.[1]

In a finite set of items, we define the concepts of *coverage* and *distribution*. *Coverage* refers to the degree to which $I(D)$ has sampled or covered the set $T \cup D$. We mean to use coverage in an intuitive way here, referring roughly to the amount of data exploration that a user has made on a dataset. Coverage is related to the notion of a cover for a set. The cover for $T \cup D$ is a collection of sets whose union contains $T \cup D$ as a subset. In terms of interactions, the cover for $T \cup D$ is the union of all sets of interactions $I(D)$ possible in the analytic process. In information visualization, the concept of coverage has been studied as a means to encourage users to explore more data [38, 48, 61, 115, 145] as well as inform users of collaborators' explorations [3, 72]. The concept of coverage is motivated by the desire to ensure that the full extent of the data is considered, even if it represents an outlier or otherwise lesser portion of the distribution of data.

Alternatively, the concept of distribution is motivated by the desire to ensure that the user's interactions with the data are proportional to the actual dispersion of the data. *Distribution*

[1] We note that in most non-streaming visual analytic systems, T and D, as well as $T \cup D$ are finite; streaming data systems have the potential for countably infinite dataset sizes, but we leave consideration of those sets to later work.

refers to the dispersion of the set of interactions $I(D)$. Distribution differs from coverage in that it accounts for repeated interactions rather than considering only the binary notion of set membership. For a set of interactions, the probability frequency function over the dimension of interest for $I(D)$ defines the shape of the dispersion of the data with which the user has interacted.

Key to our present interest in modeling evolving behavior as people interact with systems is that we can track the events in $I(D)$ that are created by the user over the course of an analytics session. We propose that by tracking these events as a Markov chain over the state space $T \cup D$, we can define metrics characterizing $I(D)$ in ways that reflect information gathering and decision-making processes. When compared to a baseline, these proposed metrics will enable us to assess when behavior differs from the baseline in meaningful ways. In the present work, we focus on meaningful deviations that might reflect cognitive biases. Further, for each metric, we define the bias value $0 \leq b \leq 1$, where higher values indicate more prominent indicators of bias, and lower values indicate less prominent indicators of bias.

For our preliminary metrics, we assume a simple baseline model of independent, equally likely interactions with any data point. At any given time, the probability of interacting with data point d_i on step $k+1$ is $P[d_{i,k+1}|d_{j,k}] = 1/N$, meaning that each next interaction does not depend on the current interaction or the interaction history. A sequence of interactions in $I(D)$ thus forms a regular Markov chain, with the data points representing the states in the chain with transition probability matrix $P = \left[\frac{1}{N}\right]$. Figure 5.2 illustrates the Markov chain resulting from four interactions with a scatterplot. The sequence of actions taken by the user was: (1) hover over point d_1; (2) hover over point d_2; (3) hover over point d_3; and (4) click on point d_3. The resulting Markov chain given in set notation is $\{\{hover, d_1\}, \{hover, d_2\}, \{hover, d_3\}, \{click, d_3\}\}$. The green trajectory over Fig. 5.2a–d illustrates the sequence of interaction events as a movement through a state space, with the growing list of {interaction, data point} dyads forming the set $I(D)$ for this user. The dashed red arrows show the unbiased baseline model, where a transition from the current (green) point to every other point, including self-transition, is equally likely.

While the assumption of uniformity is naïve, it is intended to be only a preliminary point of comparison. It allows us to establish the metrics while making few assumptions about what unbiased behavioral indicators look like, because they are likely domain and interface dependent. However, we note that the Markov chain approach allows us to flexibly swap out the transition probability matrix without altering the computation of the proposed metrics themselves. In fact, we revisit this baseline assumption further in Chap. 5.4.

5.2.2 Preliminary Metrics

We hypothesize that when cognitive bias is present, it should manifest in particular patterns of interaction with the data. In this section, we propose six preliminary metrics for detecting

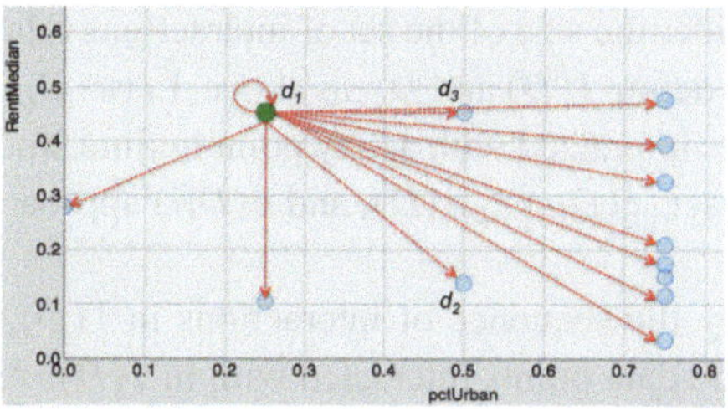

(a) Interaction 1: hover over point d_1; resulting Markov chain $\{\{hover, d_1\}\}$

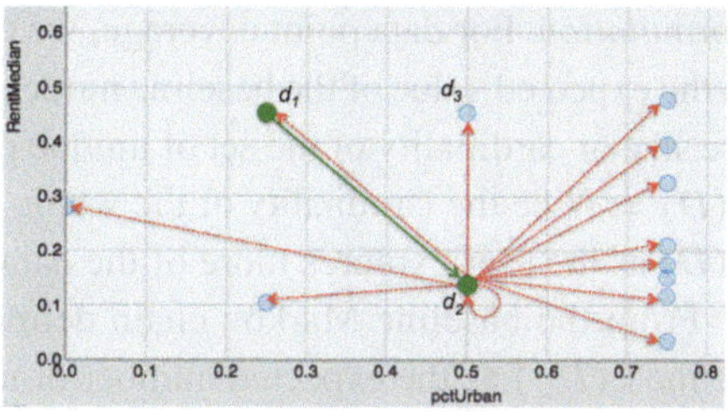

(b) Interaction 2: hover on point d_2; resulting Markov chain $\{\{hover, d_1\}, \{hover, d_2\}\}$

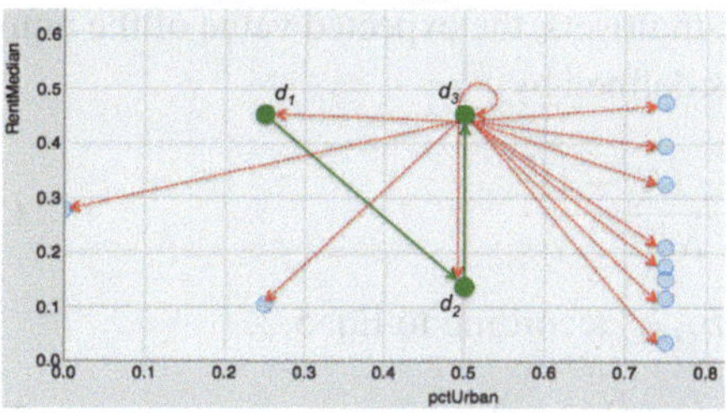

(c) Interaction 3: hover over point d_3; resulting Markov chain $\{\{hover, d_1\}, \{hover, d_2\}, \{hover, d_3\}\}$

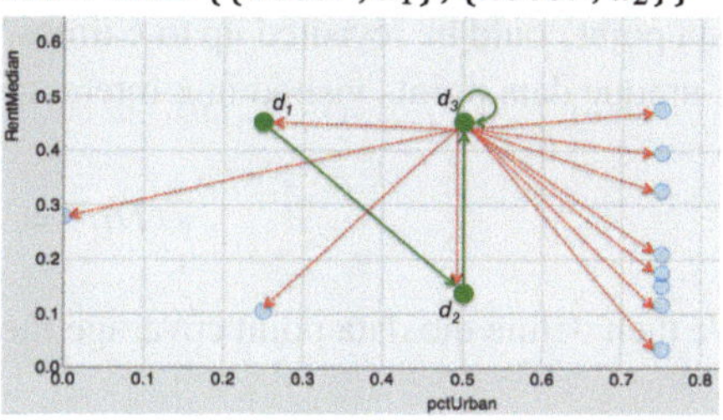

(d) Interaction 4: click on point d_3; resulting Markov chain $\{\{hover, d_1\}, \{hover, d_2\}, \{hover, d_3\}, \{click, d_3\}\}$

Fig. 5.2 The Markov chain formed by the first four interactions with a scatterplot, superimposed on top of a visualization for illustrative purposes. The set of {interaction, data point} combinations constitutes the states of the Markov chain. Subsequent interactions are conceptualized as the transitions between the states. A green point indicates a data point that has been interacted with. The red arrows indicate possible transitions from the current state

behavioral indicators of bias based on a user's interactions. We quantify behavioral indicators and define the expected values derived from the Markov chain baseline model. For each metric, we give a description, the mathematical formulation, and an example use with a type of bias from Table 5.1.

Data Point Coverage

Description. The data point coverage metric is an ordinal measure of the user's attention to the data points in the dataset. In particular, it measures the amount of the dataset with which the user has interacted compared to the expected amount. In an unbiased exploration of the entire available data, the metric decreases over time as the user interacts with more of the dataset. Of course, early in the analysis, fewer data points will have been interacted with than later in the analysis, so we must account for the number of possible interactions. So the question for the metric with respect to bias is: Is there a time in the process where the data point coverage is much smaller than would be predicted by the unbiased baseline model?

Formulation. For data point coverage, we consider the size of the set of interactions relative to the expected value of the baseline model. We define $I(D)$ and D_U as above. Let $\kappa(D_U)$ be the size or cardinality of the set of unique points interacted with at any point in time, and let $\kappa(D) = N$ be the cardinality of the whole dataset. $\kappa(D_U) \leq \kappa(D)$, and $\kappa(D_U)$ approaches $\kappa(D)$ as the user explores more of the dataset.

From the baseline Markov chain defined by the sequence of interactions in $I(D)$, we define $\hat{\kappa}(D_U)$ as the expected number of unique data points interacted with in $I(D)$. After k interactions on a dataset, or k transitions in the Markov chain, we can define a set of k-multisets, which are the sequences of length k with N possible objects in which any single data point could be revisited up to k times. In k-multisets, the expected value of the number of unique data points visited in k interactions is defined by

$$\hat{\kappa}(D_U) = \frac{N^k - (N-1)^k}{N^{k-1}}. \tag{5.1}$$

We then define the data point coverage metric b_{DPc} according to Eq. 5.2.

$$b_{DPc} = 1 - \min\left(\frac{\kappa(D_U)}{\hat{\kappa}(D_U)}, 1\right). \tag{5.2}$$

Example. To understand how this metric might be useful in capturing behavioral indicators of bias, consider the following. An analyst may propagate her/his bias by focusing on (e.g., repeatedly interacting with) or ignoring (e.g., not interacting with) certain data points. For example, when an analyst uses the *vividness criterion* [70], s/he subconsciously relies more heavily on evidence that is vivid or personal than on evidence that is dull or impersonal. Thus, bias would be propagated through the system by interacting with only a small, vivid subset of the full set of evidence.

Data Point Distribution

Description. The data point distribution metric is a measure of bias toward repeated interactions with individual data points or subsets of the data. Here we compare the frequency function of data point interactions to a baseline uniform distribution of interactions across all D. Data point distribution aids in determining if the user is focusing their interactions unevenly across the dataset.

Formulation. We can detect this by measuring the distribution of interactions with the data points. The baseline model of independent, equally likely interactions with the data points predicts a uniform distribution of interactions. We compute the χ^2 statistic, comparing the actual number of interactions with each data point to the expected baseline uniform distribution according to Eq. 5.3.

$$\chi^2 = \sum_{n=1}^{N} \frac{(\kappa(I(d_n)) - \hat{\kappa}(I(d_n)))^2}{\hat{\kappa}(I(d_n))}. \tag{5.3}$$

Here, $\kappa(I(d_n))$ denotes the observed number of interactions with data point d_n, while $\hat{\kappa}(I(d_n))$ denotes the expected number of interactions with d_n. Derived from the regular Markov chain of interactions with $P = [1/N]$, after k interactions, $\hat{\kappa}(I(d_n)) = k/N$, equivalent to the expected number of times returning to data point d_n in k steps. The p-value is obtained from the χ^2 distribution with $N - 1$ degrees of freedom, then the metric value is defined according to Eq. 5.4.

$$b_{DPd} = 1 - p. \tag{5.4}$$

Example. To understand how this metric might be useful in capturing behavioral indicators of bias, again consider the *vividness criterion* example. When an analyst uses the *vividness criterion* [70], they subconsciously rely more heavily on evidence that is vivid or personal than they do evidence that is dull or impersonal. Consequently, when evaluating evidence and forming hypotheses, they are likely to return to those most vivid pieces of information disproportionately to their actual value as evidence. This is measurable by considering the distribution of interactions across data points.

Attribute Coverage

Description. Different from considering the way the set of interactions cover the set of data points, we can also consider the way the points in D_U cover the ranges of values for the data attributes, A. Thus, for each attribute, the attribute coverage metric measures the range of values explored by the user's interactions. It gauges whether the data interacted with by the user presents a comprehensive or narrow image of the full range of values along each dimension of the dataset. If a user interacts with data in the full range of values for a given attribute, the metric will be low; alternatively, if a user only interacts with data in a small range of the possible attribute values, the metric will be high.

Formulation. Attribute coverage is computed for each attribute separately, though a single data point interaction impacts all attributes simultaneously. Attribute coverage refers to the degree to which the user interactions have sufficiently covered the range of attribute values. For categorical attributes, we define "sufficiently covered" to mean that at least one data point has been interacted with for each value $q \in Q$ that the attribute can take. For continuous attributes, we define "sufficiently covered" by quantizing the data into Q quantiles.

Let $I(D)$ and D_U be defined as above. Let Q_{a_m} be the set of Q categorical values or quantiles for attribute a_m. We then define the attribute coverage metric for attribute $a_m \in A$, according to Eq. 5.5.

$$b_{Ac}(a_m) = 1 - \min\left(\frac{\kappa(D_{U,Q_{a_m}})}{\hat{\kappa}(D_{U,Q_{a_m}})}, 1\right) \tag{5.5}$$

where $\kappa(D_{U,Q_{a_m}})$ is the cardinality of the set of values/quantiles for attribute a_m covered by the set of unique data points with which the user has interacted. Thus, b_{Ac} is greater when the user has not interacted with data over the full range of values of a_m.

Similar to the data point coverage metric, the sequence of Q_{a_m} sampled in k interactions forms a k-multiset for attribute a_m. In k-multisets, the expected value of the number of unique attribute values visited in k interactions is defined by

$$\hat{\kappa}(D_U, Q_{a_m}) = \frac{Q_{a_m}^k - (Q_{a_m} - 1)^k}{Q_{a_m}^{k-1}}. \quad (5.6)$$

As this is computed per attribute, there will be as many b_{Ac} scores as there are attributes of the data. It is possible for a person to have broad attribute coverage of some attributes and low attribute coverage of others.

Example. Consider an analyst subject to *oversensitivity to consistency* [70]. This bias can cause the analyst to dismiss evidence that is not part of the greatest encompassing hypothesis. It may lead to fruitless pursuit of an incorrect hypothesis if alternative evidence is not weighed and considered appropriately. Thus, an analyst subject to this bias might see consistent evidence that a suspect's vehicle is black and only examine black cars. The analyst might be dismissive of different accounts that the vehicle was blue or silver and consequently neglect to properly investigate alternatives. The bias would thus cause her to only interact with a portion of the range of possible attribute values in the dataset.

Attribute Distribution

Description. The attribute distribution metric is a measure for detecting bias toward particular attributes of the data. For each attribute of the data, we compare the distribution of the data interacted with to the distribution of the full dataset.

Formulation. Define $A = \{a_1, ..., a_M\}$ as the set of attributes describing the data. For numerical attributes (e.g., car price), we compare the distribution of data that has been interacted with D_U to the distribution of the full dataset D using a Kolmogorov–Smirnov (KS) test, a nonparametric test for comparing continuous distributions. The KS statistic for attribute a_m is defined by $S_{(N,n',a_m)} = \sup_x \left|F_{D,N,a_m}(x) - F_{D_U,n',a_m}(x)\right|$, where $F_{D,N,a_m}(x)$ and $F_{D_U,n',a_m}(x)$ are the cumulative distribution functions for attribute a_m over the whole dataset and the subset of unique interaction points, respectively, $n' = \kappa(D_U)$, and sup is the supremum function. We compute the empirical p-value using the KS distribution.

When the attribute a_m is categorical (e.g., gender), we apply a χ^2 test with $\kappa(Q_{a_m})$ degrees of freedom to compare the distribution of data across the categorical values. We define the test statistic according to Eq. 5.7.

$$\chi^2 = \sum_q \frac{(\kappa(a_{m,q}) - \hat{\kappa}(a_{m,q}))^2}{\hat{\kappa}(a_{m,q})}. \quad (5.7)$$

In this case, each value of q in $a_{m,q}$ represents a different value of the categorical attribute a_m. The observed value $\kappa(a_{m,q}) = \kappa(I(D))$ where $d_n[a_m] = a_{m,q}$ represents the number of data points interacted with by the analyst that have value q for attribute a_m. The expected values $\hat{\kappa}(a_{m,q})$ are derived from the actual distributions of the attribute values.

For both numerical and categorical variables, we define the attribute distribution metric b_{Ad} for attribute a_m using the p-value for the KS-test and χ^2-test, respectively, according to Eq. 5.8.

$$b_{Ad}(a_m) = 1 - p. \tag{5.8}$$

Thus, the value of $b_{Ad}(a_m)$ increases when the distribution of attribute a_m values of data points in D_U significantly differs from the distribution of attribute a_m values in D.

Example. Consider an analyst subject to *oversensitivity to consistency* [70]. If the analyst focuses on the data that is consistent with the greatest encompassing hypothesis, the distribution of the data in D_U will likely be skewed compared to the distribution D. In the case of examining suspect vehicles, 75% of the analyst's interactions may be with black cars while only 15% of the candidate vehicles are black. Thus, this metric can capture bias along particular dimensions of the data.

Attribute Weight Coverage

Attribute weights are used in visual analytic systems implicitly or explicitly to quantify the importance of each attribute in the data toward some decision. Users often specify attribute weights by interacting with interface sliders to specify each attribute's importance. The attribute weight metrics compare the coverage and distribution of weights that each attribute has been assigned by the user or system. We define an attribute weight vector $w = [w(a_1), \ldots, w(a_M)]$ comprised of numerical weights assigned to each attribute.

Description. We can consider the way the weights in w cover the possible ranges of values for the attribute weights. Thus, for each attribute, the attribute weight coverage metric measures the range of values explored by the user interactions. It gauges whether the attribute weights identified by the user's interactions present a comprehensive or narrow image of the full range of weights for each attribute. If a given attribute has had a wide range of weights applied, the metric will be low; however, if the weight for a given attribute has not taken on a diverse set of values, the metric will be high.

Formulation. With respect to attribute weights, the notion of coverage can be determined by comparing the weights the user has assigned to each attribute to the possible range of attribute weights. Again, this form of coverage is not about the shape of the distribution of weights for each attribute. Rather, attribute weight coverage refers to the degree to which the user interactions have sufficiently covered the range of attribute weight values. We first quantize each attribute's weight into Q quantiles. We then define "sufficiently covered" to mean that at some point, the weight for attribute a_m has taken on a value in each of the Q quantiles.

Let $Q_{w_{am}}$ be the set of quantiles for the weight of attribute a_m. We then define the attribute weight coverage metric for attribute $a_m \in A$, according to Eq. 5.9.

$$b_{AWc}(a_m) = 1 - \min\left(\frac{\kappa(W_{U,Q_{am}})}{\hat{\kappa}(W_{U,Q_{am}})}, 1\right) \tag{5.9}$$

where $\kappa(W_{U,Q_{am}})$ is the cardinality of the set of weight quantiles for attribute a_m covered by the set of unique attribute weights that the user has defined. Thus, b_{AWc} is greater when the user has not defined w_{a_m} to have a diverse range of values.

Similar to the attribute coverage metric, the sequence of $Q_{w(a_m)}$ sampled in k interactions forms a k-multiset for attribute weight $w(a_m)$. In k-multisets, the expected value of the number of unique attribute weights visited in k interactions is defined by

$$\hat{\kappa}(W_{U,Q_{am}}) = \frac{Q^k_{w(a_m)} - (Q_{w(a_m)} - 1)^k}{Q^{k-1}_{w(a_m)}}. \tag{5.10}$$

Example. After a piece of evidence has been discredited, analysts should re-weight attributes in accordance with new information. However, analysts subject to *persistence of impressions based on discredited evidence* [70] will likely continue to rely on the same weighting of attributes throughout their investigation. The bias would thus influence the analyst to examine a smaller part of the range of attribute weights.

Attribute Weight Distribution

Description. The attribute weight distribution metric detects bias toward particular weightings of data attributes. For each data attribute, we compare the distribution of the changes in attribute weight to a baseline exponential distribution of changes in weight.

Formulation. The attribute weight distribution metric is based on the distribution $F(\Delta w(a_m))$ of the amount of change in an attribute weight between two interactions at times τ_i and τ_j, $\Delta w(a_m) = w_{\tau_i}(a_m) - w_{\tau_j}(a_m)$. The baseline assumption is that users will be more likely to make small changes (e.g., $\Delta w(a_m)$ close to 0) to the weight of an attribute than they are to make large changes. In the present, we assume a baseline exponential distribution, $f_{\hat{\Delta}}(x) = \lambda e^{-\lambda x}$, with $\lambda = 1$. We compare the two distributions using a KS test. The KS statistic for the weight of attribute a_m is defined by $S_{(\Delta w(a_m))} = \sup_x \left|F_{\Delta w(a_m)}(x) - F_{\hat{\Delta} w(a_m)}(x)\right|$, where $F_{\hat{\Delta} w(a_m)}(x) = (1 - e^{-x})$. We then define the attribute weight distribution metric b_{AWd} for attribute a_m using the p-value for the KS test, according to Eq. 5.11.

$$b_{AWd}(a_m) = 1 - p. \tag{5.11}$$

Thus, $b_{AWd}(a_m)$ increases when the distribution of weights for attribute a_m is far from the expected exponential distribution.

Example. As with the attribute weight coverage metric, consider the example of the *persistence of impressions based on discredited evidence* [70]. After a piece of evidence has been discredited, the analyst is likely to change the attribute weights very little if at all. Thus, the tail of the distribution representing large changes in attribute weights would be smaller than the expected distribution.

5.2.3 Example Application

In this section, we present an example of how the proposed bias metrics might be incorporated into a visual analytic system.

The System. InterAxis [78] is an exploratory visual analytic system that allows users to steer scatterplot axes by interacting directly with data points. The user can interact with the data by

- hovering over a point to see details,
- dragging a point into a bin along either axis, and
- dragging an attribute bar to change its assigned weight.

Dragging points into the bins on the high or low side of the axes represents the user's semantic distinction between the data in the bins (e.g., things they like versus things they do not like, things that are important versus things that are unimportant, etc). The system then computes an attribute weight vector based on the difference between the two sets of examples for each axis. The attribute weight vector is bound to the axis, and the data points on the scatterplot are moved to their respective locations along the axis, with items semantically similar to the respective examples appearing on either side of the axis. Figure 5.3 shows the InterAxis interface with the addition of the bias metrics, described below. The original paper by Kim et al. [78] contains further details on the interface and underlying model. Herein, we use the Cars dataset [68].

Interactions. We modified the system by incorporating a custom Javascript library used to log interactions and compute the proposed metrics based on those interactions. For InterAxis, we track the interactions with data points supported by the system: hovers and drags. We also track interactions with the attribute weight vectors for each axis: when users select an attribute to bind to the axis, directly modify the weights by clicking and dragging, or when the axis is recomputed based on dragging points into the bins. The custom library uses Jerzy [108], a statistics library for Javascript, to derive probabilities from the KS test.

Bias Metrics. The six bias metrics are recomputed after each interaction. The level of bias computed for each metric is a number between 0 and 1, encoded as the width of the bar in the metric visualization (seen on the lower right of each image in Fig. 5.3). For the

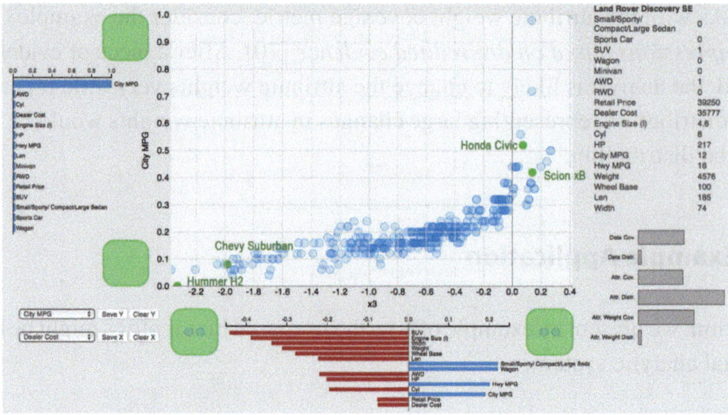

(a) Sofia's exploration of the cars in the dataset. She has dragged cars she likes to the bin on the high side of the X axis and cars she does not like to the bin on the low side of the X axis.

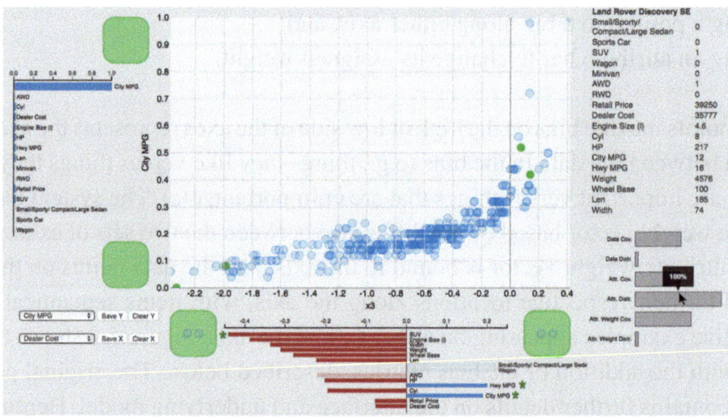

(b) The Attribute Distribution Metric. On hover, the opacity of the bars on the axes encode the metric value for each attribute. Here, Sofia notices the dark bars indicating bias along the dimensions of SUV and city and highway MPG.

Fig. 5.3 A depiction of InterAxis throughout the usage scenario. InterAxis is a system that allows users to define custom scatterplot axes using dimension reduction by interacting with data points. The proposed bias metrics have been integrated into InterAxis in the bar visualization on the lower right portion of the interface

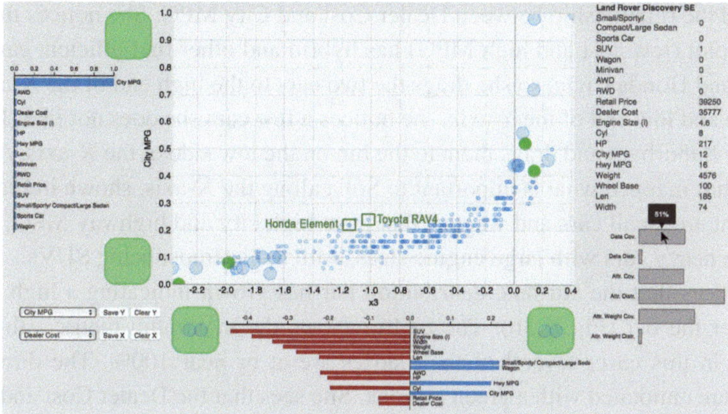

(c) The Data Point Coverage Metric. On hover, the radii of the points Sofia interacted with are increased, and the radii of the points not interacted with are decreased. Sofia notices that she has not interacted with any cars except on the extreme ends of the axes. After further exploration, she ultimately chooses Toyota RAV4 and Honda Element, unexpectedly in the middle of the X axis, to test drive.

Fig. 5.3 (continued)

attribute and attribute weight metrics, the bar encodes the maximum metric value over all the attributes. Hovering on the bars reveals a tooltip showing the percentage value for each metric, shown in Fig. 5.3b, c. Additional details are encoded in visual channels not otherwise used by the system to facilitate easier interpretation of the metrics. Hovering over the bar for *data point coverage* changes the sizes of the data points. Data points in D_U are given a larger radius, while data points not in D_U are given a smaller radius. Hovering over the bar for *data point distribution* similarly changes the sizes of the data points. Points with the most interactions are given the largest radius, while points with the fewest interactions are given the smallest radius. Hovering over the bars for the attribute and attribute weight metrics (*attribute coverage*, *attribute distribution*, *attribute weight coverage*, and *attribute weight distribution*) encodes the metric value for individual attributes as the opacity of the bars along the axes. High opacity represents a higher bias metric result. Thus, darker bars correspond to dimensions along which the bias is greater. Some of these views can be seen in Fig. 5.3.

Usage Scenario. Sofia is interested in exploring potential cars to purchase, using the Cars dataset [68]. She first changes the scatterplot axes to Dealer Cost (X-axis) and City MPG (Y-axis) to reflect her two most important criteria. She hovers over several data points to understand the relationship between Dealer Cost and City MPG. She notices the top left of the scatterplot (low cost and high MPG) has hybrid and other fuel-efficient cars. She likes Scion xB and Honda Civic, so she drags the two cars to the high end of the X-axis. Toward the middle and low end of the Y-axis, she notices a few cars she does not like (Hummer H2 and Chevy Suburban) and drags them to the bin on the low side of the X-axis. At this point, InterAxis has mapped what is important to Sofia along the X-axis, shown in Fig. 5.3a. Cars on the right are small cars and wagons that have high city and highway MPG, and cars on the left are heavy cars with large engine sizes, with the leftmost being SUVs.

She notices that the attribute distribution bar has grown indicating a high bias, so she hovers over the bar (Fig. 5.3b). The tooltip shows the maximum metric value across all attributes; in this case, several of the metrics are at or near 100%. The dimensions she examines are annotated with a green asterisk. She sees that the Dealer Cost and Retail Cost attribute bars have high opacity indicating a bias along those dimensions. Sofia acknowledges this was intentional; she is on a budget, so she interacted primarily with inexpensive cars. She also notices that the SUV attribute bar is dark. This was intentional too; she wants cars that have higher fuel economy. Next, Sofia hovers over the data point coverage bar (Fig. 5.3c). She notices she has interacted with many cars on the extreme ends of the X-axis but not much in between. Curious what types of vehicles lie in between, Sofia hovers over several cars in the mid-range of the X-axis. She sees several mid-sized cars and a few small crossover SUVs. She had previously dismissed SUVs, because she thought they had poor fuel economy.

The bias metrics computed on Sofia's interaction sequences were visualized in the interface, allowing Sofia to gain an awareness of her analytic process and biases. Her preconceived notion that SUVs have poor fuel economy led her to initially dismiss an entire class of vehicles. However, the visual characterization of her analytic process through the bias metrics shed light onto her oversight. She ultimately selects two cars of interest: Toyota RAV4 and Honda Element (annotated in Fig. 5.3c) to test drive.

While this scenario provides an illustrative example, the ideas generalize to other exploratory or decision-making tasks. Cognitive bias impacts people's behaviors in ways that can be described and quantified from their interactions. How to best present the metric information to the user is a fascinating area of future work that we discuss further in subsequent chapters.

5.3 Research: Validating Interaction Metrics—An Exploratory Analysis with Anchoring Bias

After defining theoretical foundations for bias metrics in visual analytics in the previous section, we next seek validation that the metrics are able to pick up on analysts' biased behavior. Do the bias metrics give a signal when users are engaged in analytic processes known to be influenced by bias? Further, the theoretical formulation of the metrics leaves many open questions regarding the implementation details of the bias metrics in a visual analytic tool, including (1) which interactions are used in each computation?, (2) how often to compute the metrics?, and so on. We explore these questions in the context of anchoring bias, which describes the tendency for people to rely too heavily on initial information when making a decision [46].

Originally published at the Interact conference in 2019 [133], this section presents the results of a formative study to examine how bias can be observed in users' interactive behaviors through the lens of the bias metrics, while simultaneously refining our understanding of how to apply the interactive bias metrics in a real scenario. Our goal is to leverage a well-known and highly studied form of bias (anchoring [46, 56]) to influence participants' analysis process in a controlled and predictable way. Specifically, we presented users with one of two alternative task framings, each intended to encourage users to anchor on different attributes of the data. We analyzed the bias metrics over the duration of each session, where participants were tasked with categorizing basketball players by position using an interactive visual analytic tool. While the ultimate goal of the metrics is real-time interpretation and mixed-initiative adaptation, the present work collected full interaction sequences of metrics for post hoc analysis only, to ensure the metrics can capture bias and to elucidate how to put the metrics into practice. Our analysis suggests anchoring bias elicited by task framing can be observed in users' interactive behavior through the lens of the computational bias metrics.

5.3.1 Methodology

We conducted a formative study to explore the ways that anchoring bias manifests in interactive behavior during visual data analysis, specifically through the lens of the bias metrics, described in the previous section. Anchoring bias describes the tendency for people to rely too heavily on initial information when making a decision [46]. In the analytic process, this tendency leads people to preferentially weight some information and systematically neglect other information, often leading to poorly informed decisions. In order to induce such anchoring bias on participants in this study, we utilized framing effects. Framing describes the manner in which a choice is presented to people, including the language used, the context, and the nature of the information displayed [128, 129]. Framing has been found to strongly shape decision-making [123], as people tend to take more risks under negative framing conditions. The way that information, task goals, or context are introduced to people has a strong impact on how they will conduct their analysis. By using framing effects to induce anchoring bias, we are able to evaluate how a well-known bias manifests in user interaction patterns for a visual data exploration and classification task.

Participants in the study were tasked with categorizing a dataset of basketball players. Using the visual analytic tool InterAxis [78], shown in Fig. 5.4, users were instructed to explore the dataset and label 100 anonymized basketball players according to the position

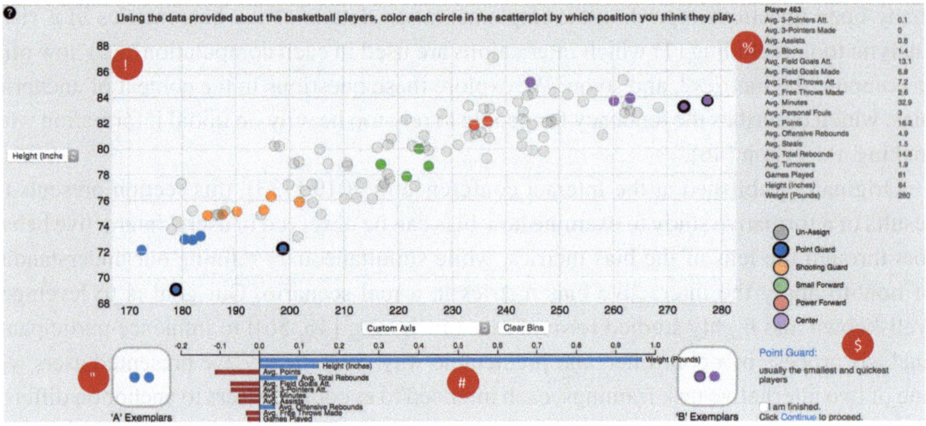

Fig. 5.4 A modified version of the system InterAxis[78], the interface used by participants to complete the task of categorizing basketball players. See text for more details

they play. Participants were randomly assigned to one of two different framing conditions, each of which described the five different positions in basketball using different attributes. The goal of using two different conditions was to anchor participants on those specific attributes during their analysis (see Table 5.3). Given the predictable ways anchoring bias influences decision-making, this study explores the research question: *can the bias metrics be used to characterize interactive behavioral differences when people are anchored on different attributes of the data?*

InterAxis

Participants used the same scatterplot-based visual analytics tool described in Sect. 5.2 to categorize basketball players by their position (Fig. 5.4). Pilot studies led us to modify the InterAxis user interface from its presentation in [78] for ease of use in the study. Changes include: the y-axis custom axis options were removed; the color scheme was changed, data point colors were changed to reflect participants' labels; options for saving the plot settings were removed; and experiment control options (e.g., Un-Assign label option, Continue button) were added. The data from the pilot was only for testing and feedback on our protocol and not included in the results.

The scatterplot view shows each of 100 basketball players represented as a circle (Fig. 5.4a). Hovering on a circle reveals details about that player (Fig. 5.4b). Data points can be dragged from the scatterplot into the bins on either side of the x-axis (Fig. 5.4c). The system, in response, will compute a custom axis using a linear dimension reduction technique. The result is a set of attribute weights that represents the differences between the points in the bin on the high end of the axis and the bin on the low end of the axis. The attribute weights are visualized as bars along the axis (Fig. 5.4d). The bars can also be interacted with by click and drag to directly manipulate the weights that make up the custom

Table 5.3 Position descriptions used in the two framing conditions. *Size* condition participants were expected to rely more heavily on size-related attributes (Height and Weight). *Role* condition participants were expected to rely more heavily on the role-related attributes called out in the description

Position	Size condition	Role condition
Center (C)	Typically the *largest* players on the team	Responsible for protecting the basket, resulting in lots of *blocks*
Power Forward (PF)	Typically of *medium-large* size and stature	Typically spends most time near the basket, resulting in lots of *rebounds*
Small Forward (SF)	Typically of *medium* size and stature	Typically a strong defender with lots of *steals*
Shooting Guard (SG)	Typically of *small-medium* size and stature	Typically attempts many shots, especially long-ranged shots (i.e., *3-pointers*)
Point Guard (PG)	Usually the *smallest* and quickest players	Skilled at passing and dribbling; primarily responsible for distributing the ball to other players resulting in many *assists*

axis. Participants can read a description of each position by clicking on the colored circles below the detail panel (Fig. 5.4e). With one of the positions selected, the user can then label players as the selected position by clicking on the points in the scatterplot.

We chose to use InterAxis for the study due to the system's highly interactive nature—to encourage users to explore and interact with the data, since the bias metrics ultimately rely on user interactions. Further, InterAxis allows users to browse data points and attributes, in addition to using an analytic model consisting of weighted attributes to project the data. This allows us to use the full set of bias metrics.

Analytic Task and Framing Conditions

Studies of anchoring bias within the cognitive science community rely on highly controlled experiments to isolate a cognitive phenomenon. However, in interactive visual data analysis, cognitive processes are often much more complex than can be captured from such experiments. Hence we sought a task with enough complexity to simulate decision-making within a realistic analysis scenario while maintaining tractable experimental conditions.

There are many tasks associated with performing data analysis in a visual analytic tool, such as ranking, clustering, or categorizing data [43, 118]. What bias looks like can be quite different across these tasks; hence, for this study we narrowed our scope to focus on categorization-based analysis. We found through pilot studies that categorizing basketball players was a sufficiently challenging task that led users to interact with the visual analytics tool for approximately 30 minutes. This provided a balance of task complexity and study tractability.

We asked participants to categorize a set of 100 basketball players by their positions based on their stats using the InterAxis visual analytic tool [78], shown in Fig. 5.4. The study dataset was a subset derived from a dataset of professional (NBA) basketball players[2] whose names and team affiliations were removed. After filtering out less active players (whose statistical attributes were too small to be informative), we randomly selected 20 players for each of five positions: Center (C), Power Forward (PF), Small Forward (SF), Shooting Guard (SG), and Point Guard (PG) for a total of 100 players. Each player had data for the following stats: 3-Pointers Attempted, 3-Pointers Made, Assists, Blocks, Field Goals Attempted, Field Goals Made, Free Throws Attempted, Free Throws Made, Minutes, Personal Fouls, Points, Offensive Rebounds, Steals, Total Rebounds, Turnovers, Games Played, Height (Inches), and Weight (Pounds).

Participants were assigned to one of two conditions. The two conditions differed in the descriptions provided for the five positions (C, PF, SF, SG, and PG). In the *Size* condition, the descriptions are based on physical attributes (Height and Weight) of players. In the *Role* condition, positions were described with respect to their typical role on the court and performance statistics. These descriptions were based on analysis of the distributions of attributes for each position as well as descriptions of the positions recognized by NBA. Table 5.3 shows the text used to describe the positions in each condition.

Participants

Ten participants (4 female, mean age 25.5 ± 2.7 years) were recruited from a large university. All but one participant had experience playing basketball, and six participants watched at least a few (NCAA, NBA, WNBA) games per season. The one participant who never played basketball watches it regularly. All participants were at least moderately familiar with information visualization. Participants were randomly assigned to either the Size or Role condition.

Procedure

Participants began by providing informed consent, completing a demographic questionnaire, and viewing a 5-minute video describing the task and demonstrating use of the InterAxis to complete the task. The demonstration used different position descriptions than the study. Participants then completed the main task, using InterAxis to categorize 100 basketball players into one of five positions. There were no time limits for completing the task. After completing the task, participants were administered a post-study questionnaire about their experience. Participants were compensated with a $10 Starbucks gift card.

Throughout the task, an investigator observed the participant's interactions, taking notes of each participant's strategies (e.g., labeling points the user was most confident about,

[2] http://stats.nba.com/

relying on a particular attribute for their categorizations, using nearby data points to visually categorize an uncertain point, etc). Participants were encouraged to ask questions as needed regarding the interface, the underlying algorithmic transformations, or the meaning of an attribute. As needed, the investigator would prompt the participant to clarify their strategy to understand when shifts in strategy occur. The investigator did not reveal information about the underlying distribution of positions in the dataset or additional attributes that might be used to help categorize players.

Timestamped logs of the users' interactions were automatically recorded, including interactions with data points (labeling, hovering to reveal details, and dragging to axis bins), interactions with axes (selecting a new attribute for an axis, dragging to adjust attribute weights, and recomputing attribute weights based on interactions with the bins), and interactions with position descriptions (clicking to reveal a description and double clicking to de-select a position description). The interaction logs capture the input data for the bias metrics.

5.3.2 Verifying Anchoring Effects

To see how the bias metrics are able to characterize anchoring bias, we first analyzed how framing more generally impacted user behaviors. We verified that the task framings induced an anchoring bias by analyzing behavioral differences between participants in the two conditions. This is an important first step to ensure that the signal detected by the bias metrics during the analysis process can be attributed to anchoring bias. The results of this analysis indicate that participants relied heavily on the attributes used in their respective condition groups.

The two framing conditions, Size and Role, were designed to bias participants in a controlled way. Our prediction is that participants will anchor, or rely more heavily, on the attributes in the particular descriptions used in their condition (Table 5.3). We first compared the frequencies of attributes selected for the axes in the scatterplot between the two framing conditions. We predicted that participants in the Size condition would select the Height or Weight attributes on the axes more than participants in the Role condition. Likewise, we predicted that participants in the Role condition would select the other framed attributes (Blocks, Rebounds, Steals, 3-Pointers, or Assists) on the axes more than participants in the Size condition.

Figure 5.5 shows the results of this analysis. Each boxplot shows the number of times the given attribute was selected on the axis for participants in the Role condition (left) and the Size condition (right). Larger separation of mean and quartile values suggests that the framing condition impacted the given attribute, while significantly overlapping boxplots suggest little or no difference between the framing conditions for that attribute. The boxplots reveal that some attribute axis selections show clear differences supporting our predictions (e.g., Height, Blocks, Offensive Rebounds, Steals, and 3-Pointers Attempted), while others

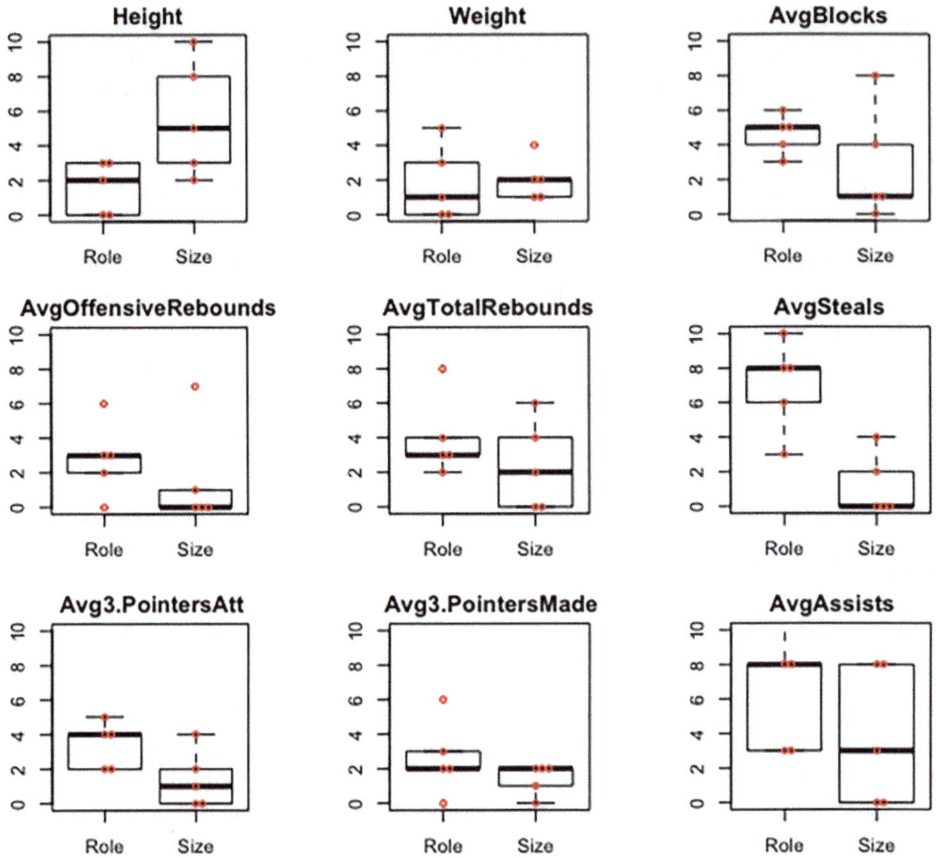

Fig. 5.5 Boxplots of number of attribute interactions via axis manipulation in InterAxis. The median is indicated by the thick middle line, the inner quartiles within the box, and the outer quartiles the whisker bars. The red dots indicate the sum of observations for each participant (rather than outliers as in traditional boxplots)

are less clearly affected (e.g., Weight, Total Rebounds, and 3-Pointers Made). These results suggest that the participants from the two groups anchored on the attributes described in the respective framing conditions.

We also characterized if participants showed different patterns in their use of the position labels between conditions. That is, anchoring bias could influence their tendency to use one label more than others, in addition to selecting the attributes differently. In categorization models, using one label more than another is often measured with a parameter called bias, referring to the bias for some categories over others. We fit the Generalized Context Model (GCM) to the categorization identification-confusion matrices [100, 101]. The GCM defines the probability of assigning players S_i from group i to position label R_j as $P[R_j|S_i] =$

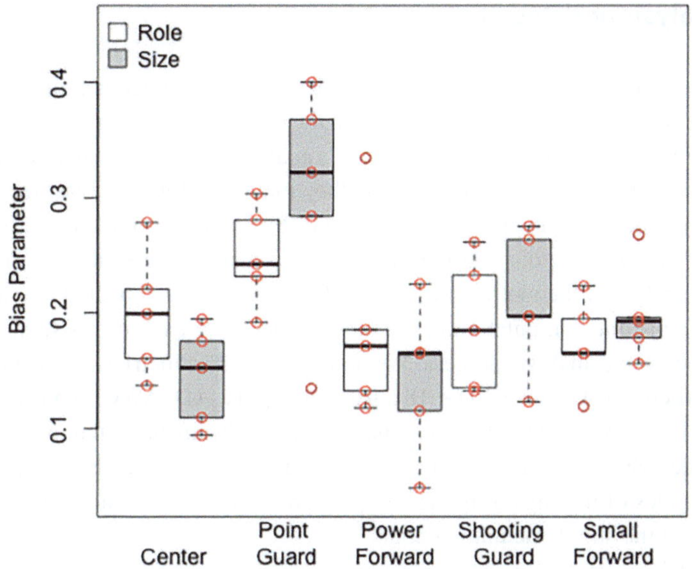

Fig. 5.6 Box plots of the GCM bias parameters estimated for each position for the Role and Size conditions. Red points represent the individual participant values

$\frac{\beta_j \eta_{i,j}}{\sum_{k \in K} \beta_k \eta_{i,k}}$, where $0 < \beta_j < 0$ is the bias for label j, subject to $\sum_j \beta_j = 1$, and $\eta_{i,j} > 0$ is the similarity of player i to all the other members of position j. We fit GCM parameters with a Bayesian MCMC estimate of 10,000 runs, holding attention weights constant, and β_j was given a uniform(0,1) prior. The mean β_j parameter estimates for each category and group are plotted in Fig. 5.6. If participants are unbiased in assigning position labels then we would expect all $\beta_j = 0.2$. Figure 5.6 shows β_j are around 0.2 for most positions. The Size condition has less variability in the β_j values, but there is evidence in that condition for a higher β_j for labeling Point Guard. The Role condition produced more variability in category bias, with most participants showing a bias toward assigning the label Point Guard over the other positions. Center and Power Forward show lower bias parameters, suggesting Role condition participants were less likely to assign those as labels to players.

Together, these results confirm that the Role and Size conditions influenced the overall categorization behaviors in ways consistent with our intended manipulations. With this overall evidence for shifting biases and attribute selection patterns, we turn to exploring the interaction patterns to see how these biases are reflected in our novel bias metrics. We note that the critical evaluation of the metrics herein relies on within-subject analysis, because the ultimate application will be quantifying the bias of a single user while performing visual analytic tasks. Consequently, our between-condition analyses remain qualitative as we compare the within-subject metric patterns.

5.3.3 Analysis and Results

We analyzed the user study data with the high-level goal of understanding how anchoring bias manifests in participants' behavior through the lens of the bias metrics. The bias metrics provide us with the ability to characterize a user's analytic process in real time by quantifying aspects of their interaction patterns in which they may be exhibiting bias. In particular, we analyze the bias metrics from the granularity of (1) the sequences of [0, 1] metric values over time and (2) where in the distribution of the data user interactions deviate from expected behavior. To analyze if the metrics can capture bias, we use the collected interaction logs to simulate the real-time computation of the bias metrics after each user's session in order to avoid influencing the analysis process. We note that the bias metrics create 74 unique time series per participant (DPC + DPD + 18 attributes x {AC, AD, AWC, AWD}). In the scope of this work, we narrow the focus of our discussion to only attributes that were referenced in the position descriptions to analyze if they picked up on the induced bias. We discuss a few selected examples of findings from the computed bias metrics. Visualizations of all metrics can be found in the supplemental materials.[3]

Participants' accuracy for categorizing players averaged 53% ($SD = 18\%$) over the course of 33.6 minutes ($SD = 14$ min). Some interactions were filtered out to reduce noise in the bias metric computations. Namely, hovers less than 100 ms were removed as likely "incidental" interactions performed unintentionally while navigating the mouse cursor to a different part of the interface. Because hovering in the interface shows a data point's details, particularly short hovers were likely not intentional interactions by the user to get information. Participants performed an average of 1647 interactions ($SD = 710$), which filtered down to an average of 791 non-incidental interactions ($SD = 300$). For additional discussion on which interactions are included in the bias metric computations, see Sect. 5.3.4.

Metrics Over Interaction Sequences

After every interaction, each bias metric computation results in a value between [0,1] quantifying the level of bias at that time. Computed over time, the metrics produce a sequence of values quantifying the level of bias throughout the analysis process which can be visualized as a time series. One way to test whether bias metrics capture anchoring is to look for differences in these sequences between the two conditions.

We hypothesized that the attributes explicitly described in each condition (Height and Weight for the *Size* condition; Blocks, Rebounds, Steals, 3-Pointers, and Assists for the *Role* condition) will have higher metric values in the associated condition than in the other. For example, we expect the time series of AD values for Assists for Role condition participants to be higher than the values for Size condition participants. To address this question, we visualized the time series for each of the 74 metrics.

[3] https://github.com/gtvalab/bias-framing

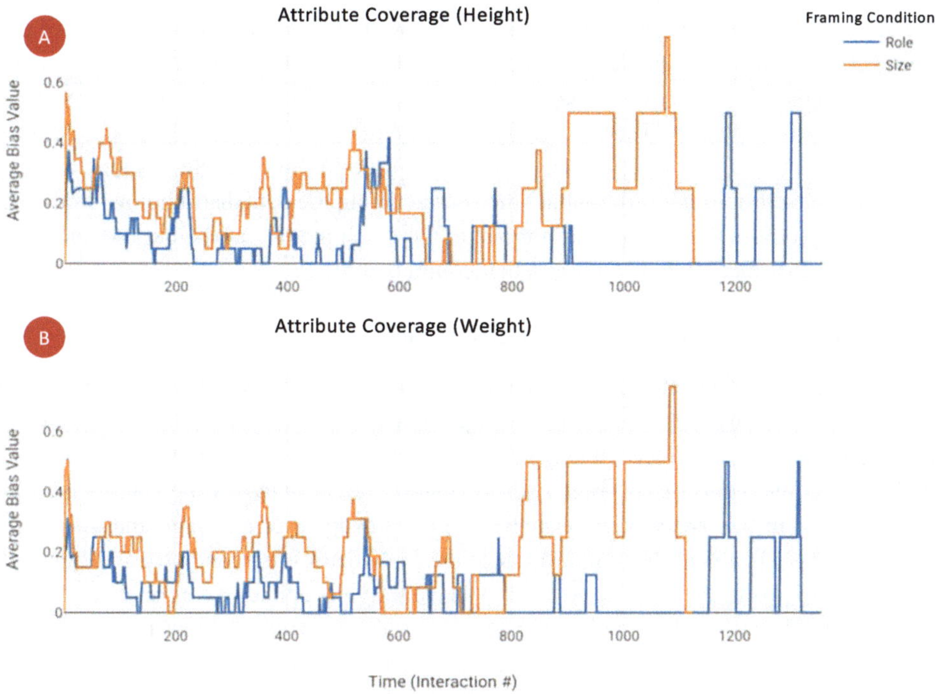

Fig. 5.7 A visualization of the average Attribute Coverage (AC) metric for the attributes (A) Height and (B) Weight for participants in each condition. Size condition participants (in orange) tended to have higher AC bias for Height and Weight than Role condition participants (in blue), consistent with our predictions

Figure 5.7 shows the AC metric for (A) the Height attribute and for (B) the Weight attribute. The blue line represents the AC metric time series averaged over all Role condition participants. The orange line represents the AC metric time series averaged over all Size condition participants. Visual examination of Fig. 5.7 finds that Size condition participants tended to have higher peaks (metric values closer to 1) and longer peaks (over greater spans of time) in the AC bias metric for the Height and Weight attributes than Role condition participants, consistent with the framing.

This visual trend can also be observed by comparing bias values averaged over the full interaction sequence for participants in each condition. Size condition participants had an average value of $M_{\text{Size}} = 0.2211$ ($SD = 0.066$) for the *Height* AC metric compared to $M_{\text{Role}} = 0.0952$ ($SD = 0.016$). Similarly, for the Weight AC metric, the Size condition participants had an average value of $M_{\text{Size}} = 0.2120$ ($SD = 0.098$) compared to the Role condition participants $M_{\text{Role}} = 0.0849$ ($SD = 0.042$).

This evidence supports our hypothesis; however, not all metrics show a discernible difference between the two conditions. One potential explanation for inconsistent effects is the level of granularity in the analysis. The bias metric values indicate the degree of bias; however, they do not indicate the source of the bias. For example, a user focusing mostly on particularly tall players might have the same metric value as a user focusing mostly on particularly short players. That is, simply knowing the metric value captures presence of a bias in the *coverage* or *distribution* of the data, attributes, or attribute weights; however, the number itself does not differentiate the source within the data distributions. In the next section, we address this by looking not just at the [0, 1] metric values, but the underlying coverage or distribution that comprises that computation.

Coverage and Distribution of Bias Metric Values

To compute the metric values, an intermediate step is to break down the user's interactions with data points, attributes, and attribute weights into quantiles and distributions. One way to illuminate the framing effects on user interaction patterns is to compare the metrics broken down into components of coverage and distribution rather than just the summative [0, 1] values. Thus, in this analysis we visualized this breakdown of coverage and distribution metrics using a heatmap. Note that because the bias metrics are computed independently for each participant, the color scale used to shade the cells is likewise normalized for each participant. That is, a black cell in the DPD metric for one participant may represent a different number of interactions than a black cell in the DPD metric for a different participant. The scales are defined in each plot.

Figure 5.8 shows what the metrics DPD, AD for Assists, and AWD for Assists look like for one Role condition participant. All of the metrics share a common x-axis of time, captured as the interaction number. The colored bars beneath the time represents the type of position being labeled during that time period (blue = PG, orange = SG, green = SF, red = PF, and purple = C). The shading in a particular (x, y) position represents the count of interactions that fall within the given bin at the given point in time, where darker shades represent a greater number of interactions.

In Fig. 5.8a, the y-axis shows a row for each data point to illustrate DPD. The DPD metric shows more bias toward players who are PGs while attempting to label PGs (2) than while attempting to label SFs (1), consistent with correct categorizations. This can visually indicate the user's bias toward particular players based on their interactive behavior. In Fig. 5.8b, the y-axis illustrates the distribution of attribute values (AD) broken down into four quantiles. The AD metric for Average Assists shows a stronger bias toward players with a high number of Average Assists while labeling PGs (1) than while labeling Centers (2), consistent with Role framing. In Fig. 5.8c, the y-axis illustrates the breakdown of attribute weight ranges (AWD) into four quantiles. The AWD metric for Average Assists indicates a bias toward higher weighting of the attribute while labeling Point Guards (1) than while labeling Power Forwards (2). The Role condition Point Guard description is intended to influence participants to anchor on the Average Assists attribute. Hence, Fig. 5.8b, c visually captures a user's anchoring bias toward an attribute.

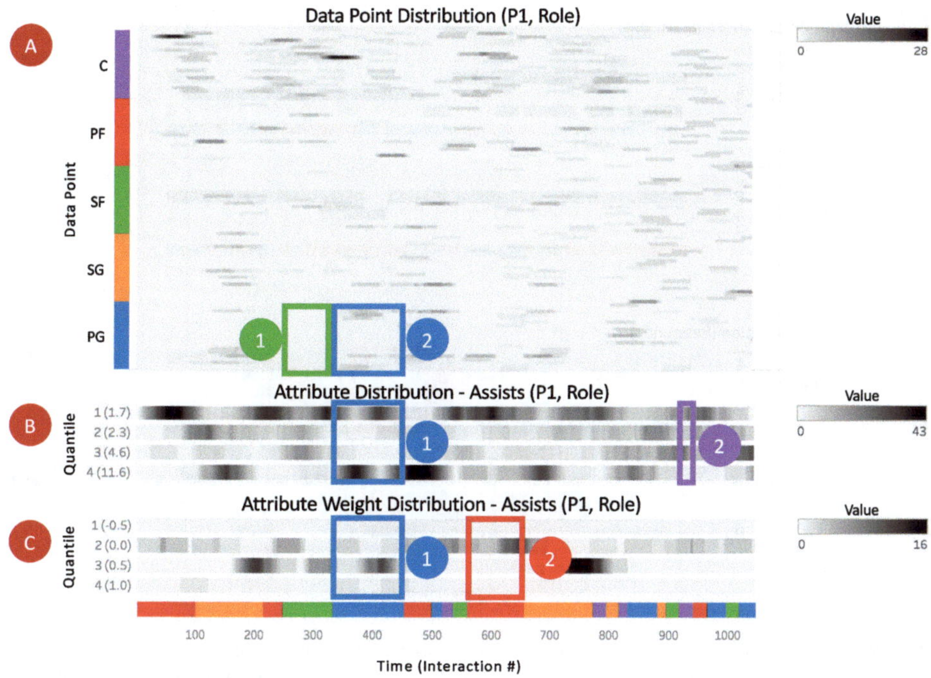

Fig. 5.8 Visualizations of three of the bias metrics for a Role condition participant: **a** the DPD metric, **b** the AD metric for Average Assists, and **c** the AWD metric for Average Assists. While labeling Point Guards (PG; blue boxes), compared to labeling other positions (SF = green boxes, C = purple boxes, PF = red boxes), the participant exhibited more bias toward PG players **a** and the Assists attribute **b** and **c** from the Role condition PG description

Figure 5.9a visually compares AWC for Height between two users from different conditions. The position descriptions used in the Size condition were designed to anchor participants on Height and Weight attributes. The Size condition participant (top) showed greater coverage of the range of attribute weights (as shown by the black bars in all four quartiles) and spent more time with a high positive weight applied to the Height attribute. Comparatively, the Role condition participant (bottom) covered less of the range of possible attribute weights and spent the vast majority of their analysis with a low weight applied to the Height attribute. We can quantify this difference using the L metric from recurrence quantification analysis [28]. L gives the average length of diagonal segments in a recurrence analysis. Applied to the metric state, larger L values reflect staying in a state longer while smaller L values reflect switching more frequently between quantiles. For the Size participant (top), $L = 14.9$ indicating more switching, and $L = 229.8$ for the Role participant (bottom), reflecting a very long time in a single quantile which is seen in Fig. 5.9a. Heatmaps for all metrics and all participants can be seen in the supplemental material.

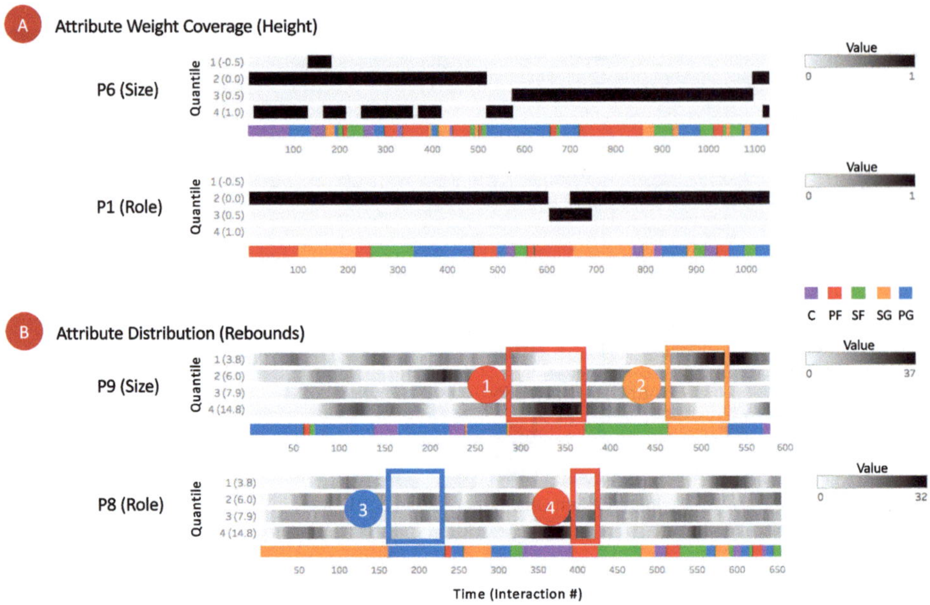

Fig. 5.9 **a** Visualization of the AWC metric. The Size condition participant (top) showed more *coverage* of the range of Height attribute weights than the Role condition participant (bottom). **b** Visualization of the AD metric for Total Rebounds. Participants focused more on upper parts of the Rebounds distribution while labeling PFs (red boxes) than other positions

Similarly, Fig. 5.9b shows how AD for Average Total Rebounds compares for one Size condition participant (top) and one Role condition participant (bottom). Role condition participants were told that PFs typically have a high number of Rebounds. While labeling PFs, both the Role condition participant (4) and the Size condition participant (1) showed interactions with greater focus toward the upper parts of the distribution (Q3 and Q4). Similarly, both the Role condition participant (3) and the Size condition participant (2) interacted with lower parts of the distribution (Q1 and Q2) while labeling other positions. While the Size condition participants were not explicitly told about the importance of Rebounds for PFs, there is a correlation between the size (Weight) of PFs and Rebounds ($r = 0.414$, $p = 0.069$), which could explain the similar patterns across the two conditions. Looking at the distribution patterns, we see both participants spent some time in all quantiles for the AD metrics. For the participant in the Size condition (top), $L = 21.2$, and for the Role condition participant (bottom), $L = 16.8$. The participants had similar L magnitudes, but the relatively larger value for Size condition participant indicates less switching between quantiles.

In summary, the task framing impacted which attributes people rely on in their interactive analysis process. These visualizations of the interactive bias metrics collectively demonstrate that anchoring bias toward particular attributes of the data can be observed in the real-time bias metrics.

5.3.4 Applying the Bias Metrics

This study constitutes the first application of the bias metrics described in the previous section, and explores some ways to interpret the resulting signals of bias. Consequently, we identified a number of challenges to consider and extracted several lessons learned in moving from theory to implementation in measuring bias through interactions. Additional sources of variability in user activities arise in the real-world analysis process that challenge theoretical assumptions. Implementation choices made early in the design process may need to adjust or adapt on the fly to accommodate unforeseen activities by the experimental participants. In this section, we present guidelines and considerations for integrating and applying the bias metrics, including a discussion on interaction design for the bias metrics, which interactions should be included in the bias metric computations, and how to interpret the metrics.

Designing for Measurement versus Usability

Fisher et al. [50] describe visual analytics as a "translational cognitive science." That is, they posit that visual analytics is a cognitive science that must travel between the often-disjointed worlds of pure science and design. In the present work, these worlds collide in the design of tools intended to measure bias from user interaction. Science motivates the interaction design to best externally reflect internal cognitive processes, while design focuses on creating a seamless and enjoyable user experience.

Designing a visualization system often involves understanding potential user needs, including things like ease-of-use, learnability, or powerful analytic capabilities. These goals each necessitate particular design decisions. Incorporating interaction-based bias metrics in an interface likewise entails its own design requirements which may conflict with other design goals. While incorporating bias computation and visualization in an interface has the potential to promote better analysis behaviors, it ultimately relies on interpreting user interactions as a meaningful capture of analytic process. Hence, the design must facilitate sufficient, meaningful, recordable interactions. In other words, the analysis process must be explicit in the interaction design of the interface.

For example, in the modification of the InterAxis [78] system for the evaluation discussed in the user study methodology section, we debated the interaction design for labeling basketball players' positions. A lasso tool could be an efficient way to label players in bulk; however, providing such a tool would make the interpretation of the interaction difficult from the perspective of the bias metrics. Further, participants would be less likely to interact with individual data points, read their individual attributes, and make a decision.

Given that the bias metrics rely on abundant interaction data, we instead decided to use single click to label data points and hover to reveal details about individual data points. This decision came at the expense of a potentially less frustrating user experience, as echoed by participants after the study. Similarly, Dimara et al. [36] conducted a study in which they had users click to locally delete individual data points on a scatterplot until a single point remained (as opposed to clicking to select the single data point in the decision). While this showed promise for mitigating bias (specifically, the attraction effect), it came at the cost of a tedious user experience.

Such trade-offs must be considered when integrating bias metrics into practical tool design. And they raise important questions for future research, such as: if the interaction design in an interface does not organically produce sufficient interaction data to measure, to what extent is it acceptable to violate user experience to achieve it?

Which Interactions to Compute On?

Incidental Interactions: The bias metrics rely on recording and computing on sequences of user interactions. Just as we must ensure that a system's interactions are designed to explicitly capture as much of the decision-making process as possible, we also need a way of knowing if some of the interactions were unintentional or lacked meaning. For example, a user may want to hover on a particular data point in the scatterplot to get details; however, due to the particular axis configuration or zoom level, the scatterplot may be overplotted. Thus, in attempting to perform a single deliberate interaction, the user might accidentally hover on several other data points along the way. These "incidental" interactions do not reflect the user's intent in any way and should thus ideally be discarded from the bias computations to remove noise. As an initial proxy for filtering out noisy incidental interactions, we ignored all hovers less than 100 ms. Some amount of noise is to be expected when leveraging user interaction as a proxy for a user's cognitive state. However, the fidelity of models can be improved by taking care to ensure, even with rough approximations, that the interactions computed on reflect a meaningful capture of user intent.

Interaction Windowing: The formulation of metrics for characterizing bias based on prior user interactions do not inform us *when* to compute the metrics or *how many prior interactions* should be computed on. In our study, we experimented with three different techniques for scoping the metric computations.

Our first approach was to compute the bias metrics after every interaction and use the *full interaction history* for every computation. Next, we tried a *rolling window* of the previous n interactions around each current interaction. The window size n then introduced another variable whose value can lead to potentially very different results. We experimented with window sizes ranging from 25 to 100 previous interactions. Lastly, we tried using key *decision points*, where the bias metrics could be computed using all of the interactions that occurred since the last decision point. We computed two variations of this: (1) using each data point label as a decision point and (2) using the activation of a position (Fig. 5.4e) as a decision point. Generalizing this windowing technique, however, requires that decision points be known, which may not be the case depending on the task and interface.

Each of these windowing techniques gives a slightly different perspective on the user's bias. For example, using the full interaction history can shed light on long-standing biases throughout the user's analytic process, while using a rolling window can capture more short-lived biases. Alternatively, using only the interactions between key decision points can be used to characterize bias in a user's interactions associated with individual decisions. As we did not know what strategies people might use, we captured short-lived biases using a *rolling window*, size $n = 50$, computed after each interaction.

Interpreting the Bias Metrics. The bias metrics are formulated such that a value $b \in [0, 1]$ is produced, where 0 indicates no bias and 1 indicates high bias. While an objective characterization of bias, the value b itself is not actionable. That is, the bias value alone does not provide sufficient detail to a user to facilitate effective reflection and correction of their behavior. For example, a user might have a high bias value for the Height AD metric. This could be due to the user focusing unevenly on short players, on tall players, or on *any* part of the distribution.

To draw actionable conclusions from the bias metric values, it is important to provide additional information to the user, specifically about where in the data or the distribution the user's interactions depart from the objective expectation. In the evaluation results, we showed one potential solution, which visualizes the *coverage* and *distribution* of interactions across data points, attributes, and attribute weights as a heatmap (Figs. 5.8–5.9). Combining both the [0,1] bias values along with the *coverage* and *distribution* that comprises the bias value computation might be ideal in some situations. For example, the [0,1] bias values could be used by automated techniques to select the most concerning dimension(s) in the user's behavior. Then, using the *coverage* and *distribution* information, systems can visualize the source of bias as the imbalance between the unbiased baseline behavior and the user's interactions.

5.4 Research: Refining Interactive Bias Metrics

We defined bias metrics and validated that they can be used to characterize anchoring bias from user interactions in a scatterplot. Now, we seek to revisit one of the fundamental assumptions made in the theoretical formulation of the metrics: that at any point in the analysis, the user is equally likely to interact with any data point. We refine the metrics in this chapter by trying to better understand what users' unbiased behavior looks like.

The baseline of unbiased behavior was theorized as a Markov model, where each combination of {data point, interaction type} constitutes a unique state. However, "unbiased behavior" was initially suggested to be represented as equal probabilities between all states in the Markov model. This assumes randomness in the user's interactive behavior, which we posit is an unreasonable assumption for most tasks and interfaces. Hence, we experimentally challenge the assumption of equal probabilities of interactions by exploring people's actual interaction sequences as they analyze data.

To refine the metrics, we replicate the study conducted in the previous section, again using anchoring bias with the basketball player categorization task. Anchoring bias describes the tendency for people to rely too heavily on initial information when making a decision [46].

However, while in the previous study we characterized anchoring bias by analyzing the *differences* in user behavior between two task framing conditions, in this study, we characterize unbiased behavior by analyzing the *commonalities* in user behavior between the two task framing conditions.

From recorded interactions, we derive a Markov model representing users' observed interactive behavior across two bias conditions. Our analysis indicates that, rather than equal probabilities of all interactions, people's interactions can be better modeled roughly based on the proximity of data points. That is, people are more likely to interact with nearby data points than those that are far away. This work was originally published at IEEE VIS in 2019 [134].

5.4.1 Experiment Methodology

We conducted a study to explore the assumptions of "unbiased" behavior in the bias metrics. In the original formulation of the bias metrics [137] and subsequent experiment [133], users' interaction sequences were compared to unbiased behavior defined by *equal probabilities for all interactions*. However, we believe this assumption is likely ill-fit for most tasks and interfaces. In recent work, researchers constructed a theoretical Markov model based on size of data points (pixel area on the screen) as an approximation for probability of interaction [30]. We are motivated by such work to create a more precise model of unbiased user behavior based on experimental observations.

We hypothesize that *proximity* can be used to better model user behavior. That is, people will be more likely to interact with nearby data points than far away data points, by starting with what they know (the initial anchoring information) and expanding their analysis, analogous to local exploration of graphs [105]. To test this hypothesis, we replicated the experiment conducted in Chap. 5.3, summarized again below, but refocused data analysis to examine probabilities of interaction sequences. Participants were randomly assigned to one of two task framing conditions, designed to *anchor* them on specific attributes of the dataset. They were tasked to utilize all of the data to categorize 100 anonymized basketball players by position (Center, Power Forward, Small Forward, Shooting Guard, or Point Guard) using InterAxis [78] (Fig. 5.4). To our knowledge, there is no known way to explore truly "unbiased" or perfectly neutral user behavior. Users will be impacted by the framing of the task, prior biases and experiences, etc. Hence, we approximate unbiased behavior by examining the commonalities between two groups of participants who are biased in a controlled way.

InterAxis

Participants utilized the same scatterplot-based visualization tool, InterAxis [78], used in the experiment in Chap. 5.3 [133].

Analytic Task and Framing Conditions

As in the previous study [133], we likewise focus on the task of data categorization. Participants were tasked to categorize 100 anonymized NBA basketball players[4], 20 players for each of the five positions: Center (C), Power Forward (PF), Small Forward (SF), Shooting Guard (SG), and Point Guard (PG). Participants were not shown the name or team of the players, but were given the following statistics: 3-Pointers Attempted, 3-Pointers Made, Assists, Blocks, Field Goals Attempted, Field Goals Made, Free Throws Attempted, Free Throws Made, Minutes, Personal Fouls, Points, Offensive Rebounds, Steals, Total Rebounds, Turnovers, Games Played, Height (Inches), and Weight (Pounds).

Participants were randomly assigned to one of two conditions. In each condition, we manipulated task framing [129] to impact users' analysis in a controlled way. The two sets of position descriptions in the task were designed to *anchor* participants on a specific set of attributes or statistics in the data (Fig. 5.4e). Participants in the *Size* condition were shown descriptions of the five positions that used statistics about their physical size (i.e., Height and Weight), while participants in the *Role* condition were shown descriptions that used statistics associated with their typical role on the court. For full experimental details, including the specific language used in each framing condition, as well as additional analyses, please refer to supplemental materials.[5]

Participants

We recruited 13 participants to complete our study (7 in the Size condition). Eligible participants completed a screening questionnaire to demonstrate sufficient background knowledge about the domain (basketball) and visualization literacy (scatterplot interpretation) [12, 87]. There was no compensation to participants in the study.

Procedure

The procedure for this experiment followed the same as in [133], with differences detailed below. Participants provided informed consent and completed two questionnaires (demographic and interface usability). They were shown videos demonstrating how to use InterAxis. Different from the procedure in [133], participants in this study were given the opportunity to get accustomed to the interface for 5 minutes with a small dataset of 15 cars to be categorized by type (as either sedan, SUV, or sports car); they were also shown a refresher video on basketball positions. The main task took approximately 15–20 minutes, during which interactions in the interface were logged. Different from [133], we collected one additional piece of information in the interaction logs to aid our analysis: the locations of all data points at the time of each interaction. In total, the experiment took about 45 minutes.

[4] http://stats.nba.com/

[5] https://github.com/gtvalab/bias-markov.

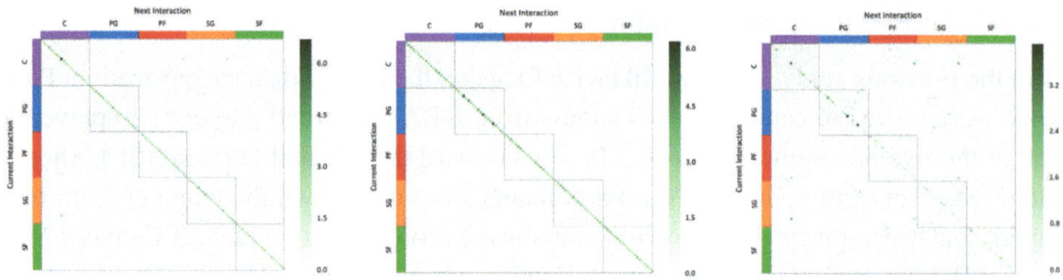

(a) Role Condition (b) Size Condition (c) Size - Role

Fig. 5.10 Aggregate probability transition matrices by condition. Rows (current interaction) and columns (next interaction) represent each of 100 basketball players, grouped by position. The highlighted squares along the diagonals indicate subsequent interactions with the same player position. Darker squares indicate higher probabilities

5.4.2 Data Analysis and Results

For simplicity in an initial model, we aggregated all interaction types (click, hover, drag) with a data point into a single Markov state. Next, we filtered out some interactions. Hovers and drags less than 100ms were likely accidental interactions [97], while the user passed from one intentional point to the next; so we removed those interactions. Participants performed, on average, 1043 interactions ($SD = 390$) which filtered down to an average of 527 interactions ($SD = 148$). Participants had an average categorization accuracy of 54% ($SD = 12\%$). Two participants (P12 and P13) did not label all 100 players in the scatterplot. They categorized $\frac{89}{100}$ and $\frac{97}{100}$, respectively. Next we describe and visualize the probabilities resulting from our analysis.

Comparing Conditions

Figure 5.10 shows aggregate matrices representing the probability of interacting with subsequent players in the scatterplot. Rows indicate the "current" interaction, and columns represent the "next" interaction. Hence, a cell is colored darker according to the probability of interacting first with the associated "current" player and then with the "next" player, where players in each matrix are ordered by their position. We see similar patterns across both conditions. Namely, there is a strong trend along the diagonal. That is, there is approximately a 50% chance that from a given state (player interaction), users next transition will remain in the same state (interact with the same player again), regardless of the condition (50.04% for role condition, 54.74% for size condition). The difference matrix between the two conditions is shown in Fig. 5.10c, revealing near-0 differences between most transition probabilities in the two conditions (98.5% of transition probabilities < 0.1). Collectively these results suggest similar transition probabilities between states, regardless of condition.

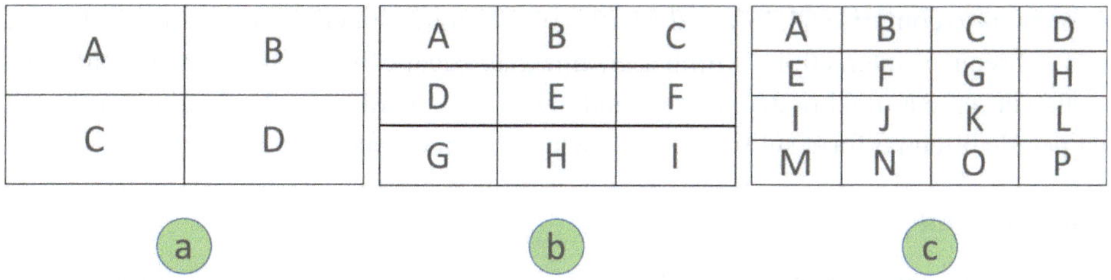

Fig. 5.11 Interactions within the scatterplot were grouped into states in the Markov model by dividing the scatterplot into **a** a 2x2 grid, **b** a 3x3 grid, and **c** a 4x4 grid

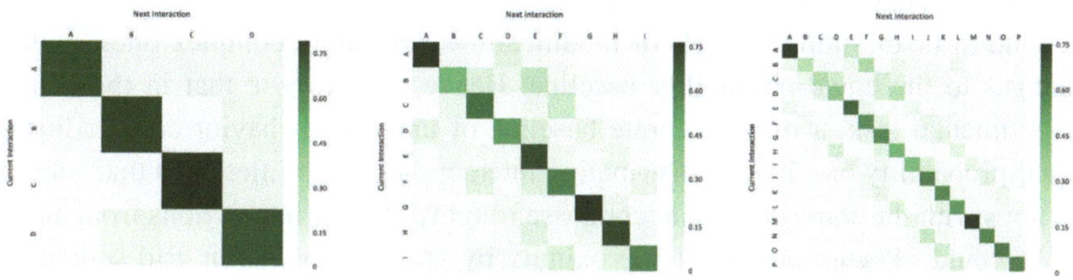

(a) $2x2$ grid (4 states) (b) $3x3$ grid (9 states) (c) $4x4$ grid (16 states)

Fig. 5.12 Aggregate probability transition matrices of all participants when Markov states are defined by grouping points in the scatterplot in a $2x2$ **a**, $3x3$ **b**, and $4x4$ **c** grid. Darker squares indicate higher probabilities

Proximity Analysis

In this analysis, we wanted to approximate the probability of interacting with visually nearby data points. To do so, we defined new Markov states by dividing the scatterplot into equal size grids (Fig. 5.11): $2x2$ (4 states), $3x3$ (9 states), and $4x4$ (16 states), and assessed the probability of interacting with points within and between these fixed grid squares. We chose to use a fixed grid overlay for our analysis in order to examine proximity even when the position of individual points on the dynamic scatterplot may be changing. From the previous analysis, we know that multiple interactions with the same player are significantly more likely (e.g., hover on a player then click to label). Hence, in this analysis, we remove subsequent interactions with the same player to see if interactions with different basketball players tend to still follow trends of proximity. Furthermore, we observe no significant difference between conditions, so here we present results aggregated for all 13 participants. Figure 5.12 shows the results of this analysis. We observe the hypothesized pattern of proximity: users are more likely to interact with other data points within the same grid square (i.e., nearby data points) than data points in different grid squares (i.e., far away data points). This is evident by the stronger colors and hence higher probabilities along the diagonal. In $Markov_{2x2}$, we find that nearby interactions (diagonal probabilities in Fig. 5.12) comprise, on average, 75.3% of subsequent interactions. Similarly, in $Markov_{3x3}$ and $Markov_{4x4}$, we find nearby interactions to comprise 64.36% and 54.29% of subsequent interactions, respectively. Apart from subsequent interactions within the same grid square (higher diagonal probabilities), we also observe a trend in $Markov_{3x3}$ and $Markov_{4x4}$ parallel to the diagonal, indicating that people often perform subsequent interactions with *adjacent* grid squares.

A New Baseline

Results of our experiment suggest that users are more likely to interact with nearby data points than far away data points when performing a categorization task with an interactive scatterplot. How do we now incorporate this information into a new probability matrix that represents a baseline of unbiased behavior?

We tend to favor simple models or modifications over more complex ones, with modest changes to the equal-probability baseline. Hence, we propose that in the context of our experimental task, a more accurate baseline of unbiased behavior could adjust from the equal-probability baseline by distributing interaction probabilities such that subsequent interactions with the *same* data point comprise roughly 50% of interactions from any given state. We could likewise account for proximity by grouping points in grid squares (as in Figs. 5.11–5.12) and defining probabilities of subsequent interactions within each grid square (nearby interactions) as at least 50% of interactions from any given state, according to the grid size chosen.

5.5 Research: Confirmation Bias—The Double-Edged Sword of Text and Visualization

People can easily become overwhelmed by data and exhibit biased behavior in visual analysis and decision-making [35, 133, 137]. One particularly insidious bias in this context is confirmation bias—the tendency to seek out and interpret information in ways that affirm pre-existing beliefs [98]. In data communication and visual analytics, this bias is especially problematic when a visualization allows for multiple interpretations, as it shapes the takeaway to align with prior beliefs, thereby hindering objective data analysis.

A common misconception among visualization readers is that visualizations present objective truths, leading them to believe that "what they see" is inherently "what is." This assumption can cause even seasoned analysts to fall prey to confirmation bias, seeing only what aligns with their beliefs and expectations and, as a result, drawing inaccurate or suboptimal conclusions. Research in psychology and visualization has shown that confirmation bias is notoriously difficult to overcome [148]. Even awareness of the bias offers little protection [35, 79].

This phenomenon can be further complicated by the complex interactions between visualizations and text. Textual annotations, which we define as text-based notes added to explain or comment on visual patterns in visualizations, can emphasize key messages to help readers spot relevant data patterns efficiently [7, 77, 121] and even shape the interpretation of the data [80, 81, 120]. In this work, we investigate the impact of a specific type of textual annotation (data facts) on confirmation bias in visual data communication. **Data facts** are natural language descriptions of data insights ranging from pointing out a single value in a visualization to complex comparisons of data distributions [119].

Widely used visualization tools, such as Tableau, Microsoft Power BI, and Google Charts [32, 49, 64], often leverage the power of automation to generate data facts from visualizations. The intent is to help data analysts extract data patterns more efficiently [67]. However, these data facts may subtly bias people [120]. For instance, consider data facts in the form of searchable natural language trends in the data (e.g., the Voder system [119]). An analyst might cherry-pick evidence that supports their hypothesis while ignoring contradictory data. Alternatively, data facts might encourage critical thinking, offering the analyst alternative perspectives that challenge pre-existing beliefs and reveal overlooked patterns. Thus, in this section, *we examine the potential of data facts to mitigate or exacerbate confirmation bias in visual analysis*. We empirically test these intuitions to systematically assess the risks and benefits of displaying data facts with visualizations. Different from the previous sections in this chapter, we measure confirmation bias not according to interaction-based metrics (Chap. 5.2), but instead as a deviation from chance in an aggregate measurement from a group of crowdsourced participants, described in more detail later. This work was originally published at ACM CHI 2025 [90].

Research Questions and Hypotheses. Across three preliminary studies and one primary experiment, we ask the research question:

RQ: How does the *presence*, *presentation style*, and *strength* of data facts impact the manifestation of confirmation bias in visual data communication?

We explore how confirmation bias may be exacerbated or mitigated by data facts (in our experimental design, we operationally refer to data facts as brief textual summaries highlighting statistical patterns in visualizations) when combined with visualizations across five hypotheses. To better understand the factors influencing confirmation bias in visual data communication, we draw on a range of psychological and data visualization research to formulate hypotheses.

First, we investigate the effect of **prior belief** on confirmation bias. Previous research in psychology has shown that confirmation bias can fluctuate in severity depending on the relationship between the user's stance and the displayed information. For example, confirmation bias is more pronounced when people hold stronger attitudes [14]. Hence, we examine whether the impact of prior belief extends to the interpretation of visual communication. This will also help us account for the effect of prior belief in later analyses to reach more generalizable conclusions on the effect of other experimental factors.

H1—Prior Belief Strength: Participants with *stronger* prior beliefs will exhibit *stronger* confirmation bias.

Similar to prior beliefs, the data topic is another contextual factor that is difficult to control when visualization researchers build tools or generate best practice guidelines for visual analysis in the real world. Data topics might interact with the strength of one's prior beliefs to consequently impact the degree of confirmation bias. For instance, political or polarizing topics in which people tend to have stronger prior beliefs may be more likely to exacerbate confirmation bias than more neutral topics [34, 132, 144].

H2—Topic: Confirmation bias will be *higher* with a more *polarizing* dataset (COVID-19 Vaccination) compared to a less polarizing dataset (Diet Choices).

With the previous two hypotheses and subsequent analysis, we can build models to control for the effect of prior belief and data topics to investigate the role of data facts in influencing confirmation bias. Data facts can either ***support*** or ***refute*** one's prior beliefs to differently affect confirmation bias. Prior research has revealed that individuals tend to perceive a visualization as less credible when the title is ***inconsistent*** with their prior beliefs compared to when the title is ***consistent*** with their beliefs [80]. We thus test how **data facts and belief alignment** impacts the manifestation of confirmation bias.

H3—Data Facts Alignment: Confirmation bias will be higher when visualizations are accompanied by data facts that ***support*** participants' prior beliefs, and lower when the visualizations are accompanied by data facts that ***refute*** participants' prior beliefs, compared to when the visualizations are shown without any data facts.

We also investigate the **presentation style** of data facts. More visualizations are generated in analytic environments with accompanying data facts [119, 141]. Further, textual data facts are often accompanied by visual annotations (in our experiment design, we operationally define visual annotations as graphical elements that emphasize a specific data point or pattern) to enhance the effectiveness of data presentation [23, 110, 112, 116]. Recent advances have even integrated auto-generated visual annotations, derived from textual descriptions of data, to streamline the annotation process [85]. These advancements in technology motivate us to investigate how representing data facts with visual annotations and textual annotations affect confirmation bias.

H4—Data Facts Presentation Style: Confirmation bias will be strongest for participants who view data visualization with ***supporting*** textual data facts only, followed by textual data facts with visual annotations, and then *No Data Facts*. For ***refuting*** data facts, confirmation bias will be lower when participants view visualizations with textual data facts only, followed by textual data facts with visual annotations.

Finally, we consider thc **strength** of the data facts themselves. Previous studies have demonstrated that captions highlighting *key visual features* in data visualizations make these features more salient [146]. Readers often interpret these captions as the primary chart takeaway [77]. Building on this, we examine whether this effect extends to data facts. We generated a set of data facts that describe different insights, referencing the taxonomies proposed by Amar et al. [1], to create a ranked list of data facts varying in persuasive **strength**. We test the effect of data facts on confirmation bias based on their strength.

Table 5.4 Results on Research Question Hypotheses. Data facts that *support* prior beliefs tend to intensify confirmation bias, while those that *refute* prior beliefs can mitigate it. Data facts describing key visual features can further amplify this effect. Additionally, visual annotations with *supporting* data facts exacerbate confirmation bias. Rows highlighted in blue indicate hypotheses that received partial or full support

Hypothesis		Support	Finding Summary
H1	Participants with *stronger* prior beliefs will exhibit *stronger* confirmation bias.	Supported	The strength of pre-existing beliefs is positively correlated with the level of exhibited confirmation bias.
H2	Confirmation bias will be *higher* with the more *polarizing* dataset (COVID-19 Vaccination) compared to the less polarizing dataset (Diet Choices).	Not Supported	The polarization level of the visualized topic does not have an effect on confirmation bias.
H3	Confirmation bias will be higher when visualizations are accompanied by data facts that *support* participants' prior beliefs, and lower when the visualizations are accompanied by data facts that *refute* participants' prior beliefs, compared to when the visualizations are shown without any data facts.	Supported	*Supporting* Data Facts lead to a higher level of confirmation bias compared to when there are *No Data Facts*. *Refuting* Data Facts result in a lower level of confirmation bias compared to when there are *No Data Facts*.
H4	Confirmation bias will be strongest for participants who view data visualization with *supporting* textual data facts only, followed by textual data facts with visual annotations, and then *No Data Facts*. For *refuting* data facts, confirmation bias will be lower when participants view visualizations with textual data facts only, followed by textual data facts with visual annotations.	Partially Supported	*Supporting* Data Facts with Visual Annotation lead to a greater level of confirmation bias compared to when there are *No Data Facts*. No significant difference was identified between *Refuting* Data Facts with Visual Annotation, *Refuting* Data Facts only, and *No Data Facts*.
H5	Confirmation bias will be higher when ambiguous visualizations are presented with *supporting* strong data facts (i.e., describing correlations) compared to *supporting* less strong data facts (i.e., describing a derived value). Confirmation bias will be lower when visualizations are presented with *refuting* correlation data facts compared to *refuting* derived value data facts.	Partially Supported	*Supporting* Correlation Data Facts lead to a higher level of confirmation bias compared to *No Data Facts*. *Refuting* Correlation Data Facts result in a lower level of confirmation bias compared to *No Data Facts*.

H5—Data Facts Strength: Confirmation bias will be higher when ambiguous visualizations are presented with *supporting stronger* data facts (i.e., describing correlations) compared to *supporting weaker* data facts (i.e., describing a derived value). Confirmation bias will be lower when visualizations are presented with *refuting* correlation data facts compared to *refuting* derived value data facts.

Summary of Findings. A summary of our findings can be found in Table 5.4. Our results showed that data facts can *exacerbate* confirmation bias when their phrasing *supports* an individual's prior beliefs, and this effect is further exacerbated when accompanied by visual annotations. Conversely, data facts can *mitigate* confirmation bias when their phrasing *refutes* an individual's prior beliefs. Our findings highlight a critical drawback of incorporating textual summaries into data visualizations: data facts can introduce bias, particularly when they align with individuals' prior beliefs. However, we also found that data facts show promise in bias mitigation when the data fact supports a refuting hypothesis. We discuss the implications of these findings for visualization design and suggest that careful consideration is needed when adding textual annotations—whether as titles or descriptions—to ensure that the interpretation remains balanced and not overly influenced by these textual elements.

5.5.1 Study Overview

In this work, we examine the effect of visual and textual annotation on confirmation bias in visualizations. This type of investigation is sensitive to contextual factors such as the visualization topic, the complexity of the data visualization, the patterns in a visualization that the data facts describe, and our participants' prior beliefs. Therefore we conducted a series of preliminary studies (PS) to identify the optimal experimental parameters for our main study, including

- *PS 1*: A range of topics with various prior belief distributions among participants.
- *PS 2*: Visualization stimuli that are *ambiguous* and, according to an individual's prior beliefs, could be selectively interpreted as supporting or refuting.
- *PS 3*: A set of data facts with different strengths to explore their impact on confirmation bias.

We conducted our main study using experimental parameters determined by the preliminary study. We provide a list of the dependent variable (DV) and independent variables (IV) we examined for the impact on confirmation bias in our main study.

- *DV*: Amount of confirmation bias.
- *IV 1 (H1)*: Prior belief strength.
- *IV 2 (H2)*: Chart topics (vaccination, dietary choices).
- *IV 3 (H3)*: Data facts alignment (supporting, refuting).
- *IV 4 (H4)*: Data facts presentation style (none, textual, textual + visual).
- *IV 5 (H5)*: Data facts strength (correlation, derived value).

5.5.2 Preliminary Studies

Participants for all preliminary studies were recruited through the crowdsourcing platform Prolific [102]. We screened for participants who are based in the United States who are fluent in English and have a study approval rate greater than 95%. Individuals who participated in any one study were ineligible to participate in subsequent ones. The study plans were reviewed and approved by the authors' university ethics board.

5.5.2.1 Preliminary Study 1: Data Topic Selection

Existing beliefs significantly impact how one processes information. The topic of a visualization can trigger varying beliefs, making the reader more or less susceptible to confirmation bias [69]. Topics unfamiliar to readers are less likely to elicit confirmation bias due to a lack of pre-existing beliefs, whereas polarizing or political topics are more likely to elicit confirmation bias as individuals' entrenched stances are less likely to move [92]. To enhance the robustness of our work, we tested various topics to identify ones associated with strong and weak prior beliefs. The main study uses topics selected from this set. This approach allows us to account for the effects of prior belief strength.

Participants. We recruited 40 participants from the crowdsourcing platform Prolific (M_{age} = 33.59, SD_{age} = 12.47; 20 Male, 18 Female, 1 Non-binary, 1 Prefer not to disclose).

Tasks and Procedure. Participants were presented statements on 16 topics (e.g., vaccine, dietary preference, government surveillance) in a random order, such as "COVID vaccination should be a mandatory policy." The full list can be found in Fig. 5.13. Following a similar setup as prior work that used belief elicitation (e.g., [148]), we asked participants to rate the "extent to which they agree with the statement" on a scale ranging from "-100 = Strongly Disagree" to "0 = Neutral" to "100 = Strongly Agree." The study ended with several optional demographic questions including their age and gender.

Measuring Topic Polarization. The data we collected presents a distribution of scores for each topic. We employed a continuous slider labeled with "Strongly Disagree" on the left, "Neutral" in the center, and "Strongly Agree" on the right. The numerical range of the slider was not visible to participants, who only saw the labels on the slider ("Strongly Disagree," "Neutral," and "Strongly Agree"). We categorized scores below 0 into the "Disagree" cluster, and those exceeding 0 in the "Agree" cluster. We aimed to select a topic associated with strong, polarizing beliefs (so that regardless of participants' stances the topic would elicit strong beliefs), as well as a neutral topic associated with weak beliefs. To this end, we computed the Cluster-Polarization Coefficient (CPC) [93] to measure the polarization of each topic. CPC captures two important features of political and social polarization: intergroup heterogeneity (BSS, which represents the variance between each cluster) and intragroup homogeneity (WSS, which represents the sum of variance within all clusters). Subsequently, we employed Equation 5.12 to compute the CPC score. The score indicates the polarization level for each topic, with 0 denoting no polarization (e.g., all participants consistently adjusted the slider toward either the left (disagree) or right (agree) side of the central point "Neutral," indicating unanimous disagreement or agreement with the statement) and 1 indicating extreme polarization (e.g., all participants in the disagree cluster moved the slider to the exact same position at the left side of the central point, while all participants in the agree cluster moved it to the exact same position at the right side of the central point). Responses with neutral beliefs were excluded when calculating CPC (3.4% of total responses), as our goal was to explore polarization on both the "Disagree" and "Agree" sides of each topic.

$$CPC = 1 - \frac{WSS}{WSS + BSS} = \frac{BSS}{WSS + BSS} \tag{5.12}$$

Results. We ranked all topics based on the computed CPC scores (full details in supplementary materials). One potential limitation of the CPC metric in our experiment setting could be that a CPC of 1 indicates extreme polarization, but does not necessarily reflect strongly held beliefs, as it could also result from participants agreeing and disagreeing uniformly on weak points. Consequently, we validated our topic selection by examining the belief distributions (as shown in Fig. 5.13) to ensure the selected topics represent both the desired polarization *and* strength of belief. The topics we ended up selecting include

- High polarization topic associated with stronger beliefs: *COVID-19 Vaccination* (CPC: 0.74).
  We selected "Vaccination" as the polarizing topic based on its highest CPC score and the concentration of strong beliefs within both the agree and disagree clusters.
- Neutral topic associated with weaker beliefs: *Diet Choices* (CPC: 0.65).
  We selected "Diet" as the neutral topic because it has the highest percentage of beliefs near the neutral point (0), with 15% of participants within the [-5, 5] range. Although two other topics, "Immigration" (CPC = 0.61) and "Work from Home" (CPC = 0.51), also have 15% of beliefs near neutral, "Diet" maintains a higher level of polarization (0.65). We chose the topic with the higher CPC score because it reduces the likelihood of a skewed belief distribution (e.g., topic 16-Pet in Fig. 5.13) among participants, which could hinder their reception to evidence that contradicts participants' common assumptions.

5.5.2.2 Preliminary Study 2: Visualization Stimuli Selection

Existing work has operationalized confirmation bias as the tendency to interpret ambiguous evidence in a way that supports one's prior beliefs [22, 29, 79]. Building on this idea, we measure confirmation bias by examining how much participants perceive an ambiguous data visualization—one that neither strongly supports nor refutes an existing belief—as supporting their own beliefs. A visualization can contain multiple patterns. Readers can draw different inferences about data depending on what patterns they pay attention to [7]. To conduct our main study, we need a set of ambiguous visual stimuli that are just complex enough to allow for various interpretations but still straightforward for participants to engage with. This preliminary study tackles this goal via two stages. First, we identify the optimal dataset size so the visualization is complex but not overwhelming. Second, we tweak the data values to generate ambiguous patterns in data that can be interpreted in competing directions. Our goal is to create stimuli with evenly divided interpretations: approximately 50% of the responses should indicate seeing a predominantly increasing trend, and the rest a decreasing trend. Using ambiguous visualizations for our study not only provides us the max amount of space to observe variance in data patterns to avoid floor and ceiling effects, but also provides a clean, conservative measure of confirmation bias as the observed confirmation bias will be less influenced by the visualization's inherent narrative. If we instead used a non-ambiguous

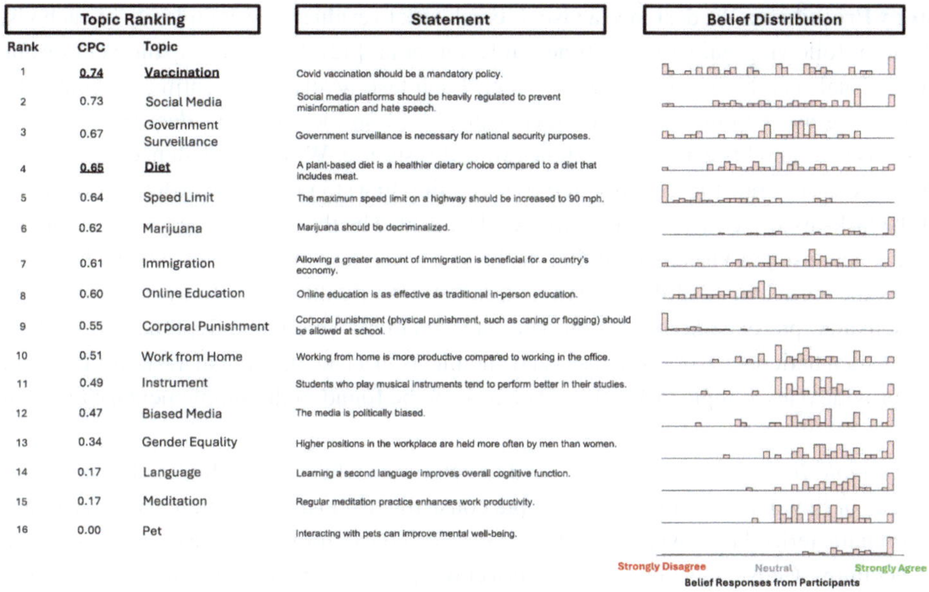

Fig. 5.13 Preliminary Study 1 Results and Belief Distributions. Participant responses were collected as slider inputs and visualized as histograms for each topic, with a bin width of 5 for each bar. The x-axis ranges from -100 to 100, with -100 representing "Strongly Disagree," 0 representing "Neutral," and 100 representing "Strongly Agree." We selected "Vaccination" as the polarizing topic, as individuals tend to hold strong prior beliefs about it, and "Diet" as the neutral topic, where people generally hold weaker beliefs

visualization that explicitly takes a stance, we would need to model its relative position to each participant's belief, introducing additional complexity and noise.

Stage 1: Determining Small Multiples Size.

Materials: Prior work suggested that bar charts depicting groups of data can elicit ambiguous takeaways [120]. Therefore we generated four types of faceted bar charts: arranged in either 3 × 3 or 5 × 5 grids, with each facet cell within these grids featuring either a single bar or two bars representing different groups. To introduce ambiguity, we balanced the number of rows displaying strictly ascending and descending trends in bar heights. We also incorporated rows that presented neutral trends (an equal number of descending and ascending trends between adjacent cells). For each design, we created two datasets to further examine whether the ambiguity persisted across different datasets. All stimuli from this stage are included in the supplemental materials.

Participants: We recruited 83 participants (M_{age} = 39.23, SD_{age} = 13.84; 44 Male, 39 Female) from Prolific. Each participant was randomly assigned to one of our visualization stimuli (3x3 single bar, 3x3 multi-bar, 5x5 single bar, or 5x5 multi-bar).

Study Procedure: We designed a visual search task to evaluate the complexity of each visualization following guidelines outlined in Elliott et al. [42]. Each participant went through three visual search tasks in a random order. Each task involved presenting a visualization and asking users to determine whether it displayed an increasing or decreasing trend in row-wise bar heights using a two-choice forced response. We counterbalanced the order of the response options (descending first or ascending first) to prevent an ordering effect. Two of the tasks used the faceted bar charts we designed. The third task served as a comprehension check, using a visualization that depicted a strict descending trend across all rows, to assess the participants' ability to extract trends from visualizations. To avoid the impact of participants' pre-existing beliefs on data interpretation, we created hypothetical scenarios about alien fictional gases (Gas Task) and chemicals in a lake (Chemical Task) for this study. We provide one example task below. The rest can be found in the supplementary materials.

- *Example Task (Gases Task)*: You need to determine which of two fictional gases, Zytharane or Skylox, is dominant in a sample consisting exclusively of these gases. Zytharane is characterized by a warming effect, where its concentration increases with rising temperatures. Conversely, Skylox exhibits a cooling effect, with its concentration decreasing as temperatures rise. This data visualization depicts the relationship between the atmospheric concentration of the gas mixture and temperature changes. To examine the dominant gas in mixture S, please analyze whether the visualization shows an overall ascending or descending trend in the atmospheric concentration as the temperature increases.

Each participant was forced to wait a minimum of 20 seconds before answering to encourage a proper examination of each visualization. After completing the three tasks, they were asked demographic questions including age, gender, and the highest level of education.

Results: As shown in Fig. 5.14a, the 5×5 multi-bar condition led to the most even split in participant response. We took a closer look at participant feedback from Prolific direct messages. Many indicated that the multi-bar visualization was confusing, and as a result, they incorrectly approached the task by comparing the two bars within the same cell for ascending or descending trends, rather than comparing the overall bar heights across the same row. Further, the 3×3 seemed was too simplistic of a visualization, such that participants overall saw either only the ascending or the descending trend (depending on the dataset). Therefore, we decided to go with the 5×5 single-bar visualization.

Stage 2: Generating Ambiguous Data.

In order to further improve our 5×5 single-bar visualization to be consistently ambiguous, we adjusted the specific data values to reach an even 50-50 split among participants in terms of reported ascending or descending trends.

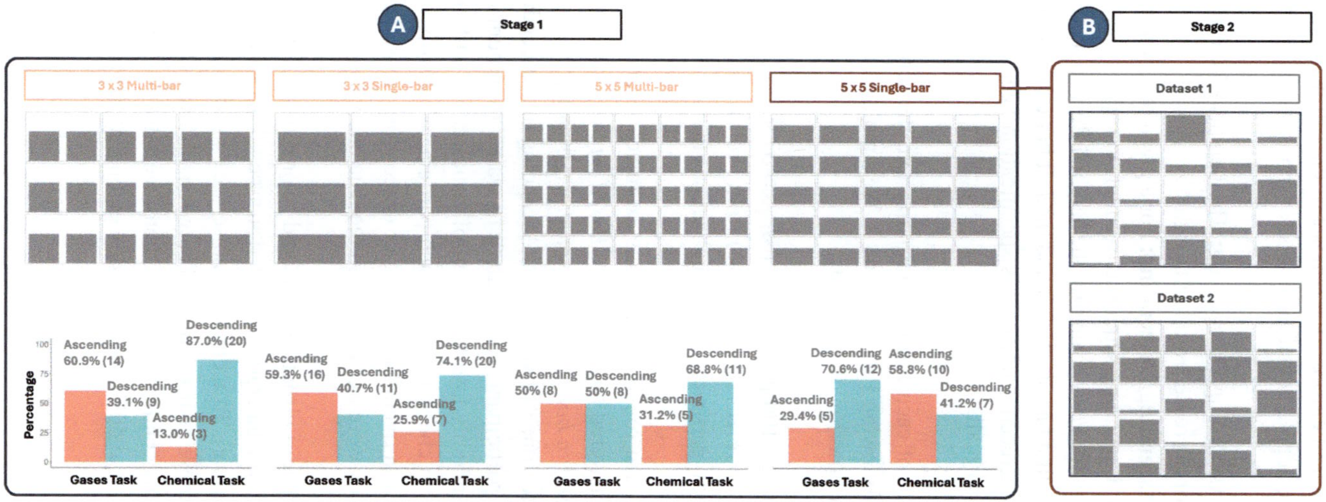

Fig. 5.14 Preliminary Study 2: Stage 1 **a** and Stage 2 **b** Conditions. In Fig. 5.14a, the depicted charts under assessed dimensions are simplified graphical representations of the stimuli, the actual depicted charts in our study include titles, axis labels, and tick marks (see supplemental materials). At the bottom of Fig. 5.14 A, the bar charts illustrate the distribution of evaluations for each stimulus dimension in Stage 1. We selected 5 × 5 single-bar stimuli, colored in brown, and chose datasets 1 and 2 from the 16 generated datasets in Stage 2 as depicted in Fig. 5.14b

Materials. We developed a Python script that simulates data values in a 5 $\times$ 5 matrix to test a range of stimuli. The script enforces the balance of the following patterns for each row to maintain ambiguity (see supplemental materials for more details):

- **Pattern 1 Strictly Ascending/Descending:** The heights of the five bars in a row progressively increase or decrease from left to right.
- **Pattern 2 Overall Ascending/Descending:** For a given row, four out of five bars follow a strict ascending or descending sequence from left to right.
- **Pattern 3 Neutral trend:** For a given row, the height difference between any symmetrical pair of bars (e.g., the first bar on the left and the last bar on the right, the second and the fourth bars) will not exceed 3 units. This specification keeps enough variability in the neutral trend to appear realistic.

To ensure the ambiguity of each resulting visualization, the script ensures there exists an equal number of rows following Patterns 1 and 2. We simulated 16 ambiguous data visualizations that satisfy our criterion of ambiguity.

Participants. We recruited 100 participants (M_{age} = 37.92, SD_{age} = 12.38, 45 Male, 53 Female, 1 Non-binary, 1 Prefer not to disclose) from Prolific and conducted two rounds of this preliminary study (50 participants, 8 visualizations in each round).

Study Procedure. In each round, participants examined eight ambiguous visualizations in random order and answered the question, *"Based on the visualization provided, please identify the general trend in bar heights as you move from left to right across each row."* Participants then selected one of two options: Ascending or Descending. To ensure a thorough evaluation, each participant was required to wait a minimum of 20 seconds before responding. We also randomly inserted a control visualization with five strictly ascending rows as a comprehension check to assess each participant's ability to reason whether the majority of rows depict an overall ascending or descending trend. Participants who failed the comprehension check were excluded from the data analysis.

Results. An ideal ambiguous visualization would be one in which participants responded in equal proportions that the visualization depicted an ascending vs. descending trend (i.e., 50% ascending, 50% descending). To account for noise, we consider stimuli sufficiently ambiguous if participants' response distributions fell between 40% and 60% ascending/descending. Five stimuli satisfied this condition. Among them, we selected the two visualizations that elicited responses closest to 50% ascending/descending, as depicted in Fig. 5.14B.

5.5.2.3 Preliminary Study 3: Strength of Data Facts

Data facts can vary in their persuasive strength [103, 120]. To account for the effect of data fact strength on confirmation bias, we conducted this study to identify data facts with strong and weak persuasiveness.

Participants. We recruited 40 participants ($M_{age} = 33.35$, $SD_{age} = 9.29$; 21 Male, 17 Female, 1 Non-binary, 1 Prefer not to disclose) from Prolific. Each participant read visualizations with ten types of data facts presented in a random order.

Materials.
Data Facts: We designed ten types of data facts aligned with the ten low-level analytic tasks delineated in framework of Amar et al., similar to the approaches by Srinivasan et al. [1, 119]. These examples are included in the supplemental materials.
Visualization: We used one of the 5 × 5 small multiples bar charts as depicted in Fig. 5.14, depicting the vaccination topic. This visual stimulus is also included in the "Example Visual Stimuli and Data Facts" panel in Fig. 5.15.

Study Procedure. After obtaining informed consent, each participant read 10 visualizations, each accompanied by a line of text describing a unique data fact, in random order. We employed the same slider settings as described in Sect. 5.5.2.1 to elicit participants' prior beliefs on vaccinations with the prompt: "Please rate the extent to which you agree with the following statement: COVID vaccination should be a mandatory policy." To measure the *confirmatory* effect of each type of data fact, we manipulated all data facts to align with participants' prior beliefs. We also included an attention check question: *"If you are paying attention, please drag the handle to the left end of the slider. This is an attention check."* For each visualization, participants evaluate it by responding to the prompt: *"How strongly do you perceive this visualization and the accompanying text to* ***support*** *the takeaway: COVID vaccination should be a mandatory policy?"* Participants responded using an interactive slider ranging from strongly refuting (-100) to strongly supporting (100). The numerical ranges for both sliders, eliciting prior beliefs and visualization evaluations, were hidden from participants.

We also included a comprehension check at the beginning of the study to ensure participants understood how to read our visualization stimuli. We asked: *"What was the mortality rate on May 22nd among partially vaccinated individuals?"* as a multiple-choice question. Participants who failed the attention check or the comprehension check were excluded from the analysis. Participants who held a neutral belief were also excluded as this makes it impossible to interpret potentially confirmatory belief updating.

Results. For each data fact, we calculated its perceived data fact strength score following Equation 5.13. The term $Response$ represents the input from each participant's interactive slider. The data fact strength score reflects the extent to which the participant perceived the visualization (which was designed to be ambiguous and neutral) to be aligned with

Example Visual Stimuli and Data Facts

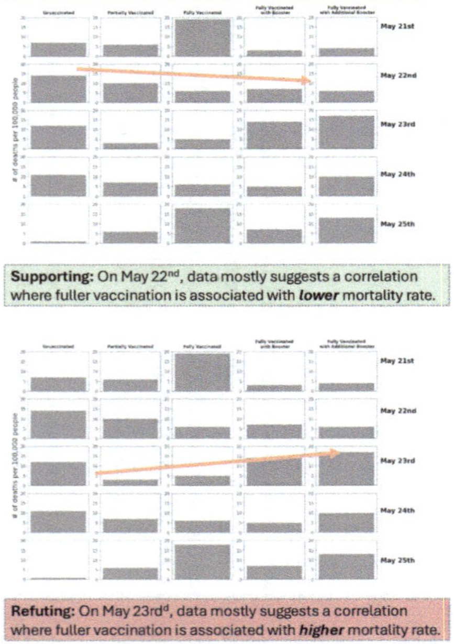

Ranking Results

Rank	Data Facts	Example	Average Strength
1	Correlation	Data mostly suggest a ***positive correlation*** where fuller status is associated with higher mortality rate	
2	Range	From May 21st to May 25th, the ***range*** of mortality rates was from 1 to 14 deaths per million among unvaccinated individuals, compared to 5 to 19 for fully vaccinated individuals	
3	Filter	Mortality rates among fully vaccinated individuals with additional booster surpassed 12 deaths per million people ***on both May 23rd and May 25th***.	
4	Cluster	From May 23rd to May 25th, ***the pattern was similar*** with mortality rates among fully vaccinated individuals with additional booster being higher than that among fully vaccinated individuals with booster.	
5	Extremum	The ***highest*** mortality rate on May 23rd was observed among fully vaccinated individuals with additional booster.	
6	Retrieve Value	The mortality rate among fully ***vaccinated individuals with Booster on May 23rd*** was 14 deaths per million people.	
7	Describe Distribution	On May 23rd, the mortality rates across all vaccination levels ***were skewed towards*** those with less complete vaccination statuses.	
8	Sort	The group order of increasing mortality rates ***from the lowest to the highest*** on May 23rd is: Partially Vaccinated < Fully Vaccinated < Unvaccinated < Fully Vaccinated with Booster < Fully Vaccinated with Additional Booster.	
9	Derived Value	The ***average*** mortality rate was 7.68 deaths per million people among all individuals from May 21st to May 25th.	
10	Outlier	On May 25th, the mortality rate among fully vaccinated individuals is ***unusually high***.	

Strongly Refute Prior Beliefs — Neutral Evidence — Strongly Support Prior Beliefs

Direction Of the Evidence

Fig. 5.15 Preliminary Study 3: Data Facts and Results. The strength of each type of data fact was assessed based on how strongly participants perceived the data visualizations and facts as supporting their prior beliefs. The mean value and 95% confidence intervals for each type of data fact are depicted in the confidence interval plot. The color coding indicates whether the mean rating of each data fact leans toward supporting or refuting participants' prior beliefs their prior beliefs. We designed the experiment such that participants only interacted with visualizations containing supportive data facts for evaluating confirmatory effects. A positive data fact strength score (+) means the participant reported the visualization to align with their prior beliefs, suggesting that the data fact biased the visualization toward those beliefs. Conversely, a negative score (-) means the participant reported the visualization to refute their beliefs, indicating that the data fact biased the visualization against those beliefs despite the data fact being designed to be supportive.

$$Data\ Facts\ Strength = \pm\,|response| \tag{5.13}$$

We calculated the average strength score for each type of data fact across all participants to produce a ranking of the strength for each data fact, as summarized in Fig. 5.15. We ultimately selected "Correlation" as the strong data fact and "Derived Value" as the weak data fact for our main study. We used "Derived Value" instead of "Outlier" as the data facts with the lowest strength because the "Outlier" condition resulted in a negative score, which could suggest unfamiliarity with the concept and potential misunderstanding. Hence we avoided using this data fact type.

5.5.3 Main Study

After the three preliminary studies, we conducted our main study to address our research questions in a pre-registered experiment[6]. Figure 5.16 shows the experimental variables and procedure. A summary of our findings is presented in Table 5.4, with more detailed analyses available in the supplemental materials.

Materials.

- *Topics*: To validate the impact of data facts across a broader range of pre-existing belief strengths, we selected two topics as determined from the Preliminary Study 1:
  - **[COVID-19 Vaccination]** A more polarizing topic where individuals tend to hold strong beliefs either in support of or opposition to mandatory COVID-19 Vaccination.
  - **[Diet Choice]** A less polarizing topic where individuals tend to have relatively more neutral beliefs on whether a plant-based diet is healthier than one that includes meat.
- *Visual Stimuli*: Based on the results of Preliminary Study 2, we selected the ambiguous 5×5 small multiples bar charts with a single bar per cell as the stimuli, with two synthetic datasets.
- *Data Facts*: Two types of data facts with varying strengths accompanies our visualizations, based on the results of Preliminary Study 3.

[6] https://aspredicted.org/N78_C7G

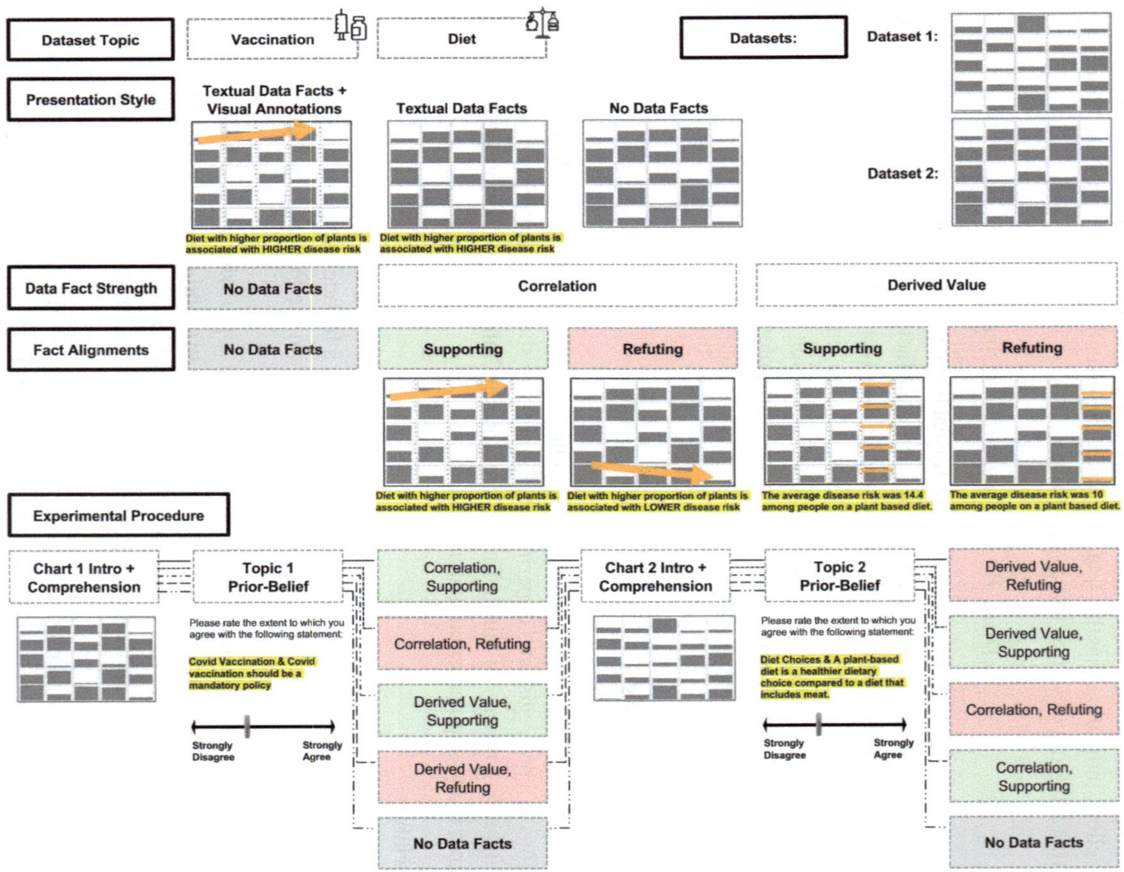

Fig. 5.16 Main Study Conditions and Procedure: [Presentation Style] is a between-subjects condition. Each participant completes 2 trials, either involving 2 data visualizations *with data facts*, where [Dataset, Dataset Topic, Data Facts Strength, Fact Alignment] are within-subject conditions, or 2 visualizations with *No Data Facts*, with [Dataset Topic] as the only within-subject condition. The varying types of strokes connecting the components in the main study procedure diagram represent the different conditions assigned to each participant

- **[Correlation]** A more compelling data fact describing the correlation between column and row variables.
- **[Derived Value]** A less compelling data fact describing the average value of certain columns.

Conditions. Participants were randomly assigned to one of three presentation styles as between-subjects conditions: textual data facts with visual annotations, textual data facts only, or *No Data Facts*. In all conditions, each participant read two visualizations, with the topic and datasets counterbalanced with the display order, as shown in Fig. 5.16. For those in the textual data facts only and the textual data facts with visual annotations conditions, each participant read both a visualization with a correlation data fact and one with a derived value data fact, one of which supports their prior belief and the other refuting. Their display order and combination are counterbalanced. For example, if the first visualization showed a supporting correlation data fact on the topic of vaccines via dataset 1, the second visualization showed a refuting derived value data fact on the topic of Diet Choices via dataset 2. For the *No Data Facts* condition, participants viewed the two ambiguous visualizations without any data facts, with the dataset topic order randomized.

Task and Procedure. Our study procedure is shown in Fig. 5.16. After completing the consent form, participants start with a warm-up activity to familiarize them with our stimuli. During the warm-up, they read chart descriptions explaining information encoded by rows and columns, followed by a comprehension check to ensure they understood how to interpret the chart. Participants who do not pass this check are excluded from our analysis.

Next, we assessed participants' pre-existing beliefs about each topic using an interactive slider ranging from "strongly disagree" (–100) to "strongly agree" (100) with the prompt: "Please rate the extent to which you agree with the following statement:" Statements related to the topics were provided as outlined in Table 5.5. Participants then read the bar charts, either with a textual data fact, with a textual data fact plus visual annotations, or with *No Data Facts*. Afterward, they were prompted to rate their level of agreement with the same statements in Table 5.5, using an interactive slider ranging from "strongly *refute*" (-100) to "strongly *support*" (100). The numerical ranges of both sliders were not visible to participants. An additional trial on the topic "Productivity while working from home" was included as an attention check and appeared randomly between the two main trials: *"If you are paying attention, please drag the handle to the left end of the slider. This is an attention check."* Participants who failed this check were excluded from our analysis. Participants who held a neutral belief were also excluded as this makes it impossible to interpret potential confirmation bias.

Table 5.5 The statement for each topic

Topic	Statement
COVID-19 vaccination	COVID vaccination should be a mandatory policy
Diet choices	A plant-based diet is a healthier dietary choice compared to a diet that includes meat

Measuring the Strength of Prior Beliefs: We measure the strength of each participant's prior belief on each topic as the absolute distance from either endpoint of the slider ("strongly disagree" and "strongly agree"), based on the following equation:

$$belief\ strength = |initial\ belief| \,. \tag{5.14}$$

In Equation 5.14, $initial\ belief$ represents the participant's response from the interactive slider. If a participant *agrees* with the statement provided, we measure the strength as the absolute distance from the right end of the slider. Conversely, if a participant *disagrees* with the statement, we measure the strength as the absolute distance from the left end of the slider.

Measuring Confirmation Bias. We evaluated how participants interpreted an ambiguous data visualization in relation to their prior beliefs on each topic to assess confirmation bias. On the slider, a value of 0 indicates a neutral stance toward the statement in the prompt. Therefore, the confirmation bias score was calculated using the formula:

$$bias\ score = \pm\,|evaluation| \,. \tag{5.15}$$

In this equation, $evaluation$ is the participant's rating of how much the ambiguous data visualization either supports or refutes the statement given in the prompt. The bias score is assigned a positive (+) sign when the evaluation of the visualization aligns with the participant's prior beliefs and a negative (-) sign when it contradicts those beliefs.

Participants. We recruited 1,080 participants from Prolific ($M_{age} = 38.79$, $SD_{age} = 12.96$; 531 Male, 537 Female, 5 Non-binary, 2 Transgender, 1 Genderqueer, 1 Genderfluid, 3 Prefer not to disclose). We screened for participants who are based in the United States who are fluent in English with study approval rate greater than 95%. Individuals who participated in preliminary studies were ineligible to participate in the main study. The study plan was reviewed and approved by the authors' university ethics board. The sample size was determined by a power analysis conducted using G*Power [47] to ensure a minimum power of 0.75 (indicating a 75% chance of detecting a true effect for each hypothesis if it exists) across all effect sizes from our hypotheses, with an alpha level of 0.05. This analysis was based on a pilot study involving 61 participants under the same settings as the main study.

5.5.4 Results

See Table 5.4 for a summary of results.

Methodology. We use a single mixed-effect linear model to test H1, H2, and H3. For H4 and H5, we separately examine the effects of supporting and refuting data facts using two linear models:

- **Situation** SN **:** Analyzing data that includes only participants who viewed visualizations with *Supporting* Data Facts and *No Data Facts*.
- **Situation** RN **:** Analyzing data that includes only participants who viewed visualizations with ***Refuting*** Data Facts and *No Data Facts*.

This separation is crucial, as we hypothesize that *supporting* data facts increase confirmation bias compared to the no data fact condition (our baseline), raising the score above zero. On the other hand, ***refuting*** data facts decrease confirmation bias compared to our baseline, lowering the score below zero. Analyzing these conditions together could obscure key differences in the distributions due to the effects evening out.

5.5.4.1 Analysis of H1, H2, and H3

We created a mixed-effect model to analyze the data. The fixed effects included

- the strength of initial belief, as calculated from Equation 5.14 (a continuous integer from 0 to 100);
- the data topic (COVID-19 Vaccination or Diet Choices);
- the data fact alignments (*Supporting*, ***Refuting***, or the reference level: *No Data Facts*).

Participant ID is the only random effect. The dependent variable is the amount of confirmation bias, as calculated from Equation 5.15. The results are summarized in Table 5.6.

H1: Prior Beliefs

As shown in Fig. 5.17, under **H1** *we found that participants' confirmation bias score is significantly positively correlated with the strength of their pre-existing beliefs, regardless of the direction of data facts*. With one unit increase in the strength of the pre-existing belief, the confirmation bias score increases by about 0.473 units. This **supports H1**.

H2: Topic

While we observe a trend that the COVID-19 Vaccination topic resulted in higher confirmation bias ($Estimated\ Marginal\ Mean(EMM) = 20.1$) compared to the diet choice topic ($EMM = 17.7$), we observe no significant effect of topic. Thus H2 **is not supported**. One

Table 5.6 Mixed-effect linear modeling results. The fourth column, "Coefficient P Value," represents the p-value obtained from the model fitting process for each category under each independent variable. The fifth column, "Variable P Value," represents the p-value derived from the Type III ANOVA for each independent variable

Independent variable	Coefficient	Std. error	t-value	Coefficient p-value	Variable p-value
Initial belief strength	0.473	0.034	13.771	< 0.001 ***	< 0.001 ***
Topic	2.414	1.867	1.293	0.196	0.196
Alignment: Refuting	-6.570	2.600	-2.527	0.012 *	Alignment: < 0.001 ***
Alignment: Supporting	7.417	2.600	2.853	0.004 **	

possible explanation for the lack of significance could be that the topical differences are captured by prior beliefs (**H1**). In particular, people hold stronger prior beliefs toward polarizing topics ($p < 0.001$, 95% CI for the mean difference in initial belief strength between diet and vaccination topics: [-18.27, -13.30]). As a result, when both prior belief strength and topic are included in the model, the topic itself might not significantly impact confirmation bias.

H3: Data Facts Alignment

Compared to the *No Data Facts* condition, participants exhibited significantly more confirmation bias when reading *Supporting Data Facts* ($EMMsup - -EMMn = 7.4$), while they exhibited significantly less confirmation bias when reading *Refuting Data Facts* ($EMMn - -EMM_{ref} = 6.6$). These results **support H3**.

5.5.4.2 Analysis of H4

We conducted the statistical analysis under both situations SN and RN, as outlined at the beginning of Sect. 5.5.4. Since this is a between-subject comparison, we constructed a linear model with the following predictors:

- the strength of initial belief (a continuous integer from 0 to 100, calculated using Equation 2);
- the topic of the dataset (COVID-19 Vaccination or Diet Choices);
- the data facts presentation style (textual data facts only, textual data facts with visual annotation, or *No Data Facts*).

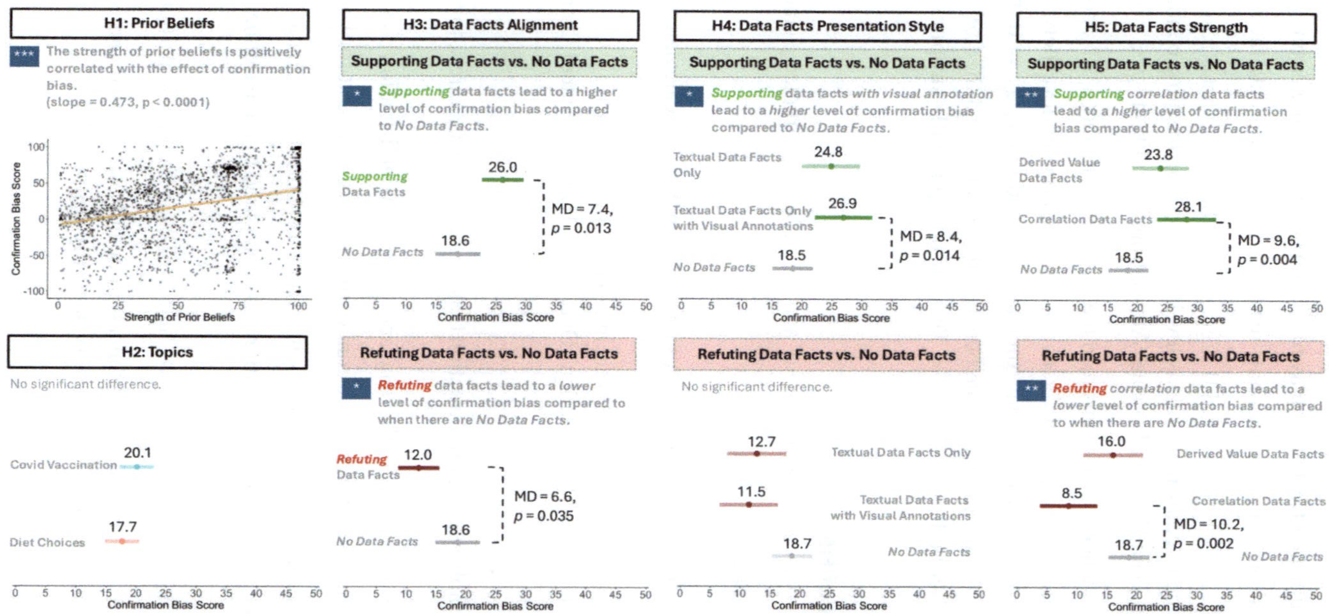

Fig. 5.17 Main Study Hypotheses and Results. The results of **H1** are illustrated as a scatterplot showing the relationship between the strength of prior beliefs and confirmation bias. The *estimated marginal means* from each linear model and the 95% confidence intervals of confirmation bias scores across different conditions in **H2–H5** are plotted as confidence interval plots. For **H2**, colors represent different topics, while for **H3, H4, and H5**, dark red indicates ***refuting*** data facts, forest green represents ***supporting*** data facts, and gray signifies no data facts

The dependent variable was the confirmation bias score. The results are summarized in Table 5.7.

We found a significant effect of display modes on confirmation bias under both SN and RN. For the situations where participants read the ***supporting*** data facts SN, post hoc analysis with Bonferroni correction [143] suggests that the significant effect is driven by the significant difference in confirmation bias levels between the baseline condition (***No Data Facts***) and [Textual Data Facts with Visual Annotations] condition ($p_{\mathrm{SN}} = 0.014$).

For the situations where participants read the ***refuting*** data facts RN, post hoc analysis suggests only a trending significant difference in confirmation bias level between the baseline and the [Textual Data Facts with Visual Annotations] condition ($p_{\mathrm{RN}} = 0.051$).

We observed no significant difference between the baseline and the [Textual Data Facts Only] condition under both SN and RN. These results **partially support H4**.

5.5.4.3 Analysis of H5

We again consider both the situations SN and RN. We constructed a linear model with the following fixed effects:

- the strength of initial belief (a continuous integer from 0 to 100, calculated using Equation 2);
- the topic of the dataset (COVID-19 vaccination or Diet Choices);
- the strength of data facts ([Correlation] Data Facts, [Derived Value] Data Facts, or ***No Data Facts***).

The dependent variable was the confirmation bias score. The results are summarized in Table 5.7.

We found a significant effect of the data fact strength on confirmation bias under both SN and RN. Post hoc analysis with Bonferroni correction suggests that for both ***Supporting*** and ***Refuting*** data facts, participants who read the [Correlation] data fact exhibited a significantly **higher** amount of confirmation bias compared to the baseline ($p_{\mathrm{SN}} = 0.004$, $p_{\mathrm{RN}} = 0.002$).

On the other hand, for both ***Supporting*** and ***Refuting*** data facts, participants who read the [Derived Value] data fact did not exhibit a significantly different amount of confirmation bias compared to the baseline. These results **partially support H5**.

Table 5.7 Fixed-effect linear modeling results. The sixth column, "Coefficient P Value," represents the p-value obtained from the model fitting process for each category under each independent variable. The seventh column, "Variable P Value," represents the p-value derived from the Type III ANOVA for each independent variable

Situation	Independent variable	Coefficient	Std. error	t-value	Coefficient p	Variable p
SN	Style: Textual Only	6.348	3.011	2.108	0.035 *	Style: 0.008 **
	Style: Textual + Visual	8.451	2.980	2.836	0.005 **	
	Strength: Correlation	9.688	3.039	3.187	0.001 **	Strength: 0.005 **
	Strength: Derived value	5.333	2.952	1.807	0.071	
RN	Style: Textual Only	–5.905	3.061	–1.929	0.054	Style: 0.0279 *
	Style: Textual + Visual	–7.229	3.029	–2.387	0.017 *	
	Strength: Correlation	–10.134	2.997	–3.382	0.001 **	Strength: 0.003 **
	Strength: Derived value	–2.694	3.086	–0.873	0.383	

5.5.5 Implications for Visualization Research

5.5.5.1 The Role of Visual Annotation

In our findings on **H4**, we were surprised to observe that *supporting* data facts might only increase confirmation bias when accompanied by visual annotations, suggesting that supporting data facts alone may not significantly bias users. This unexpected finding motivates us to reconsider the role of *visual annotations* in visualization design.

The use of annotations and embellishments in visualizations has long been debated within the visualization community. While some researchers advocate for maximizing the data–ink ratio to reduce chart junk that could distract or mislead viewers [125, 126], others argue that annotations can improve the accuracy of information recalled from visualizations [5, 71]. More recent research demonstrated that annotations can significantly influence people to draw conclusions aligned with the annotations, beyond what might naturally capture their attention [7, 120]. These insights together provide a potential explanation for our findings on **H4**: namely, visual annotations could enhance the salience of patterns described by textual data facts and thus shape how individuals interpret a visualization as *supporting* or *refuting* evidence, which ultimately affects the manifestation of confirmation bias. Therefore, designers should be aware that when visual annotations are combined with textual summaries, there is a strong potential to consolidate viewers' prior beliefs when there is alignment, making individuals more susceptible to confirmation bias.

5.5.5.2 (De-)Biased or Echoing?

The opposite effects of *supporting* and *refuting* data on confirmation bias prompt us to consider potential driving factors. Were people prompted to think more critically when they read *refuting* data facts, or did they just echo back the trends described by the data facts? Our results demonstrated that *refuting* data facts cause people's beliefs to shift 35% (see Fig. 5.17 H3 Bottom)[7], resulting in an estimated marginal mean of confirmation bias scores of 12.0, significantly greater ($p = 0.035$, 95% CI [8.59, 15.5]) than a baseline of zero (which means no confirmation bias). If people were simply echoing the perspective of the *refuting* data facts, we would expect greater belief movement in the opposite direction, with sub-zero values for the confirmation bias score. But this was not the case. In other words, people still exhibited confirmation bias (albeit less) even when viewing *refuting* data facts. This finding supports the case that *refuting* data facts help mitigate confirmation bias, rather than the case where participants just echo the trends described by the data facts. This observation also corroborates that confirmation bias is a weighting bias rather than an interpretation bias [89], where people assign higher weight to evidence that *supports* their prior beliefs than to evidence that *refutes* them.

5.5.5.3 Exacerbating or Mitigating Confirmation Bias?

Consider a viewer who holds a strongly positive belief about COVID-19 Vaccination. If they encounter a data visualization with data facts that *support* their existing belief, it could strengthen their conviction by 40% (see Fig. 5.17 H3 Top). However, if they are presented with data facts that *refute* their prior belief, it could weaken their conviction, shifting their belief by 35% in the opposite direction (see Figure 5.17 H3 Bottom)[8]. This suggests that *refuting* data facts might help reduce confirmation bias. While the distribution of bias scores with *refuting* data facts remains positive (suggesting that some confirmation bias is still

[7] 35% is calculated from the estimated marginal means in Fig. 5.17 H3 Bottom as (18.6-12)/18.6.

[8] 40% is calculated from the estimated marginal means in Figure 5.17 H3 Top as (26-18.6)/18.6, and 35% from the means in Figure 5.17 H3 Bottom as (18.6-12)/18.6.

present), it is nevertheless lower than in scenarios with *No Data Facts*, indicating a potential trend toward reducing confirmation bias with *refuting* data facts relative to the baseline level of bias. Our findings corroborate previous research in data visualization that highlights the power of text alongside data visualization to shape individuals' interpretations of data [77, 80, 120]. These findings might also indicate that providing alternative interpretations of the same visualization—especially those that challenge prior beliefs—might offer a more balanced perspective of the data and might prevent decisions based solely on *supporting* evidence.

Data Facts Consideration Our experimental findings revealed that data facts have potential both to exacerbate and mitigate confirmation bias depending on whether they are framed as *supporting* or *refuting* the viewer's prior beliefs. Realistically, as the use of large language models proliferates through more application areas, it will be hard to avoid the inevitable combinations of visualizations supported by automatically (or manually) generated textual summaries. Rather, we assert that this is an opportunity to leverage the mitigating potential of data facts in future systems. In particular, if future systems leverage belief elicitation techniques [82] prior to showing users data, systems can intelligently display refuting data facts alongside visualizations, thereby mitigating confirmation bias in interactive settings. Nevertheless, there is a clear risk that malicious actors could misuse these insights to induce a favorable bias. We encourage responsible use of these insights, not as a tool for unfair manipulation, but as a tool to promote balanced thinking and informed decision-making.

5.5.5.4 Refuting Data Facts and Counterfactuals in Visualizations

Existing work in visualizations have examined other approaches to reduce confirmation bias in visual data analysis. One such approach involves the use of counterfactuals, which refer to the mental constructs that represent *alternatives* to reality [20, 113]. When people think "if only" or "what if," they imagine how the past could have unfolded differently. This process helps reduce bias by encouraging individuals to consider alternative factors that could have led to the same outcome. In an analytical scenario where the data suggests two variables might be causally related, counterfactuals encourage users to question whether the two factors are truly causally related or if a third variable might be influencing the results [138–140]. In another example, highlighting points that have previously been examined (thereby visually distinguishing points that have/have not been previously examined) when analyzing data with visualizations encourages users to focus more on unexplored areas to potentially gain new insights [48, 135]. Further, providing people with *refuting* data facts, as we did in our study, can also prompt people to consider alternative perspectives. Thus, counterfactuals, highlighted data values, and *refuting* data facts offer users alternative perspectives that challenge their existing beliefs about the data. These insights underscore the potential de-biasing power of incorporating multiple data perspectives within a visualization. By empowering analysts to consider a wider range of alternatives for a balanced analysis, these techniques can combat over-reliance on existing beliefs and reduce confirmation bias.

5.5.5.5 Limitations

One limitation of our study is the narrow set of topics tested. We selected only one polarizing topic and one neutral topic, with their levels of polarization evaluated by a small group of 40 participants. However, there are other topics where people overwhelmingly share the same stance, such as the general affection people have for pets [2]. Future research should test a wider range of topics to further understand the generalizability of our findings.

Another limitation is the use of synthetic datasets. While these datasets allowed us maximum control over our experiment, they may not fully capture real-world scenarios relevant to the topics studied. For example, a participant who is particularly savvy in the topic may suspect the authenticity of the dataset and evaluate associated evidence in unpredictable ways. Additionally, in the present study, we operationalized confirmation bias as interpreting ambiguous information in a way that aligns with one's prior belief, and thus designed visualizations to be intentionally ambiguous. In the real world, not all visualizations would be designed to be ambiguous; some are explicitly crafted to highlight strong data patterns and convey a particular narrative. Confirmation bias would undoubtedly happen with these visualizations as well. Existing work by Xiong et al [147] has demonstrated that in these cases, visual representation formats, such as showing the data via a table or a bar, would change the strength of confirmation bias. Therefore we encourage future work to also investigate the effect of data facts and visual annotations on confirmation bias in these scenarios where the visualization stimuli are not ambiguous.

Furthermore, the real-world factors related to the statements in Table 5.5 are likely more complex than what is depicted in the data. For instance, in the COVID-19 Vaccination topic, participants might also consider factors such as mortality rates across different age groups to determine if the vaccination should be mandatory, while in the Diet Choices topic, other health metrics beyond the risk of chronic disease—such as Body Mass Index (BMI), nutrition levels, and emotional well-being—are also indicators of diet health when making comparisons. The limited factors provided in our data might introduce distortions in users' evaluations that are, to some extent, inevitable in controlled studies. Future work can incorporate more realistic datasets to generalize findings in ecologically valid settings that better reflect the complexity of real-world decision-making processes.

Finally, we also determined the strength of the data facts solely based on modulating which visual patterns they described (*PS 3*, Sect. 5.5.2.3). Further research could explore how factors such as the phrasing of the data facts, source credibility, and trustworthiness influence their perceived strength.

5.6 Research: Unmasking Dunning–Kruger Effect in Visual Reasoning and Visual Data Analysis

Imagine that two colleagues, Bob and Jane, are tasked with analyzing visualizations of model output for financial forecasting at their company. Bob, despite his limited experience in data analysis, confidently takes charge of interpreting the financial charts. He gravitates toward a line chart with smoothed trends, which he interprets as conclusive evidence of growth. His confident interpretation presents an incomplete, overly simplified view of the data, masking critical fluctuations and anomalies indicated by other visualizations of the model output. This could potentially lead to misguided strategic decisions based on an incomplete understanding of market dynamics. On the other hand, Jane, with a strong background in financial analysis, focuses on a scatterplot that details individual data points representing key financial metrics with confidence intervals around each prediction produced by the forecasting model. Her interpretation highlights variability and outliers, providing a more comprehensive view crucial for informed decision-making, suggesting a more cautious approach to future investments. Despite the precision of her analysis, Jane lacks confidence, believing that her nuanced interpretation might not be the best to base decisions on.

Both Bob and Jane exemplify a cognitive bias known as Dunning–Kruger effect (DKE). In the seminal paper titled "Unskilled and Unaware of It," Kruger and Dunning describe a phenomenon in which the people who perform the worst on various knowledge tests have an inflated perception of their abilities [83]. Bottom-quartile performers believed that their performance was above average, while those in the top quartile underestimated their performance relative to their peers [83]. This lack of realization about one's own skill reflects a metacognitive deficit, i.e., a lack of "knowing what we know" and "knowing what we don't know" [8].

DKE can have numerous consequences. In the opening example, Bob's tendency to express uninformed views likely prevents other colleagues with more fruitful perspectives from participating. This phenomenon may also affect organizations, in which the most capable people may not be the ones making decisions; instead, those with the greatest self-perceived ability (and often lesser actual skill) take precedence. Hence, the social consequences of this bias can lead to larger systemic problems [122]. Namely, DKE can lead to situations wherein true expertise may not reach the decision-making table, dominated instead by those who may be unaware of their own lack of proficiency.

We posit that DKE also can critically affect visual reasoning and judgment tasks. People with limited knowledge in the domain of interest or in visual data analysis practices may be prone to overestimating their ability to accurately reason with data supported by visualizations. The consequences of DKE in visual reasoning and judgment tasks could result in people confidently reporting on flawed analyses or drawing incorrect conclusions. Likewise, highly skilled users might underestimate their abilities, which can lead to a lack of confidence in their interpretations or decisions. This could cause second-guessing of sound analyses or potentially missing important insights or trends in the data.

In this section, we describe work that was originally published at the Workshop on TRust and EXpertise in Visualization [26] and subsequently expanded at IEEE VIS [25]. **We demonstrate that DKE can be replicated in visual reasoning and judgment across two experiments.** In the first study, DKE can be observed in participants' visual reasoning abilities using a sliding puzzle game. In the second study, DKE can be observed in participants' visual judgment through use of an interactive scatterplot to categorize data. Furthermore, our work builds on a prior body of work in interaction analysis that suggests relationships between (i) interactive behaviors and personality traits [16] and (ii) interactive behaviors and cognitive biases [133]. Specifically, we demonstrate that those who are most susceptible to DKE (i.e., have the biggest gap between their actual and perceived performance) tend to have higher values for the conscientiousness personality trait in the puzzle game (Study 1) and the car task (Study 2) and perceive they have greater domain familiarity in the car task and credit task (Study 2). In addition, those who performed at the extreme high and low ends in both studies exhibited some distinct patterns of interaction.

To our knowledge, this work is the first of its kind to address *metacognitive deficits* in visual reasoning and judgment. By demonstrating the presence of DKE in visual reasoning and judgment tasks, we provide further evidence of its pervasive impact. In addition, understanding *how* the bias manifests through interactive strategies and how it relates to personality traits paves the way toward personalized bias detection and targeted mitigation strategies. All materials and analyses are in supplemental materials: https://github.com/CAV-Lab/DKE_supplemental.git.

5.6.1 General Method

We conducted two complementary user studies that cover visual reasoning and judgment tasks. In the first study (Sect. 5.6.2), participants arranged tiles in a 15-puzzle game, and in the second study (Sect. 5.6.3), participants completed a data categorization task using an interactive scatterplot. These tasks engage spatial reasoning and pattern recognition skills [76, 142] as well as interactivity [149], which are critical for making sense of data in visualizations [149]. In this section, we outline the general method and hypotheses that are common to both studies.

5.6.1.1 Procedure

After providing informed consent, participants began with the 20-item Big Five Personality inventory [39], which produces a score from 4 to 20 for 5 personality traits: Openness, Conscientiousness, Extraversion, Agreeableness, and Neuroticism. Participants then completed the main task (15-puzzle game in Study 1; data categorization in Study 2), during which we logged their interactions with the interface. Two attention check questions were interspersed.

Participants were compensated at $10/hour based on the estimated completion time for each study.

Afterward, participants were asked to estimate their performance relative to their peers as a percentile along two dimensions: solution optimality (Study 1) or accuracy (Study 2), and reasoning ability (both). This estimation was done in a post-survey with questions such as *"how efficiently (fewer moves is better) do you think you did relative to your peers?"* A higher percentile means that the participant perceived that they performed better than their peers. Participants were also asked to rate their familiarity with the puzzle game (Study 1) or data domain (Study 2), e.g., *"To what degree do you consider yourself familiar with the puzzle game?"* on a 5-point Likert scale (1: not at all familiar, 5: extremely familiar). We asked participants about their domain familiarity after the task to prevent any influence on their confidence or perceived expertise before the task. This approach aimed to gather unbiased performance data and assess the correlation between perceived familiarity and actual/perceived performance without pre-task bias.

5.6.1.2 Hypotheses

Across the two user studies, we hypothesized that

H1 **DKE in Visualization.** We will replicate DKE within the context of visualization. Specifically, less competent individuals will overestimate their task performance relative to peers, while competent individuals will underestimate their corresponding performance percentile.

H2 **Performance and Interactions.** To our knowledge, no previous work has specifically explored whether there are indicative behaviors associated with the DKE. Hence, we hypothesize that there will be detectable differences in interactive strategies used by individuals who are more and less competent.

H3 **Interactions and Personality.** People with different personalities will display different interactive strategies.

H4 **Personality and Δ Performance.** The difference between actual and perceived performance (Δ Performance) quantifies the magnitude of metacognitive miscalibration in DKE. We hypothesize that there will be correlations between personality traits and Δ Performance.

H5 **Δ Performance and Domain.** Previous studies have demonstrated that personal familiarity with a particular event or outcome tends to boost comparative optimism [95]. Hence, we hypothesize that people's overestimation of their performance will be positively associated with their familiarity in the respective domains.

5.6.2 Study 1: Visual Reasoning with 15-Puzzle Game

We selected a puzzle game as our first task because of its relevance to spatial reasoning [31] and pattern recognition [24] that are key to reasoning with visualizations [130, 131]. Importantly, it also serves as a relatively simple task, characterized by few interactive elements, which can be a valuable starting point to first verify the relevance of DKE in visualization prior to exploring more complex interactions and tasks. The goal of this study is to (1) examine if DKE exists in the context of visual reasoning, (2) investigate users' interactive behaviors when performing the task, and (3) examine if their personality traits are indicative of this bias. To realize the three goals, we designed a pre-registered experiment.

5.6.2.1 Experimental Setup

Task and Interface. The primary view of the 15-puzzle game, as depicted in Fig. 5.18, features a 4x4 grid with 15 numbered tiles and one empty space, allowing tiles to be moved by dragging them into the empty slot (A). The goal of the puzzle is to rearrange tiles in ascending numerical order (1, 2, ..., 15) in the *least number of moves possible*. Below the board, a move counter increments by one with each move (B), allowing participants to track their total number of moves. Participants were informed at the beginning of the study that the back button was disabled to prevent reversing the move count. To guarantee comparability,

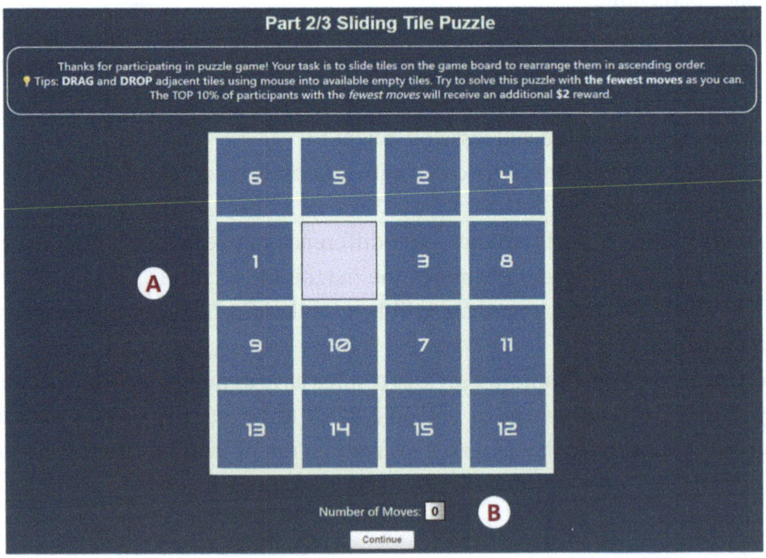

Fig. 5.18 15-puzzle interface with **a** the primary puzzle and **b** move counter. The tiles shown represent the initial board configuration used in Study 1

all participants started the game with the same initial configuration (as shown in Fig. 5.18). Users' interactions (including tiles clicked, positions they moved from/to, and time stamps of each movement) while performing the task were recorded for analysis.

Task Difficulty. Every possible state of a 4x4 board is solvable in 0 to 80 moves [18]. We conducted preliminary pilot studies to calibrate the appropriate task difficulty. These pilot studies involved a variety of puzzle sizes, ranging from the simpler 8-puzzle (3 by 3 layout) to a more complex 24-puzzle (5 by 5 layout). These trials were instrumental in gauging performance across a spectrum of difficulties, leading us to settle on the standard 15-puzzle (4 by 4 layout) configuration. We chose an initial configuration that could be solved in 10 moves, as determined by the A* algorithm [66], indicating a moderate difficulty level.

Recruitment. We conducted a power analysis of pilot data with 18 participants who completed the 15-puzzle using a similar experimental setup. Our minimum target sample size was 36 participants to obtain .8 power to detect a medium effect size of .25 at the standard $\alpha = .05$. We recruited 48 participants in total. Participants were compensated $2.50 for the study which had an estimated duration of 15 minutes (actual median completion time of 7.5 minutes). Prolific allowed participants to spend a maximum of 56 minutes on the task; no participants timed out. Participants were incentivized with an additional $1 performance bonus if they completed the puzzle in the top 10% based on lowest move count. The performance bonus was awarded to 23 people who completed the puzzle in the optimal number of 10 moves. No data were excluded from analyses due to failed attention checks in this study; however, 9 were discarded due to (i) data loss resulting in missing logs (5 people) and (ii) erroneous logs resulting from refreshing the browser which reset the task (4 people). This may lead to a skew in the data (e.g., poor performers with high move counts may be the ones who were more likely to refresh). In total *we collected and used data from 39 individuals* in our forthcoming analysis.

Participant Demographics. Among the 39 participants who completed the study, 18 identified as female, 20 identified as male, and 1 preferred not to disclose. The majority of participants (25) hold a college degree (Bachelor's, Master's, or Doctorate), while 10 have some college or Associate's degree, and 4 have a High School diploma. Participants were on average 34.87 years ($SD = 11.71$) of age. Participants rated familiarity with the solution of the puzzle game on average 2.23 out of 5 (with 33 participants reporting a familiarity level of 3 or lower).

5.6.2.2 Results

Analysis of DKE is based on a comparison of actual and perceived performance percentiles, particularly for the top quartile performers and bottom quartile performers. 23 participants completed the task in the optimal number of 10 moves. Thus for the forthcoming analysis on DKE, we defined quartiles of actual task performance primarily based on minimal move

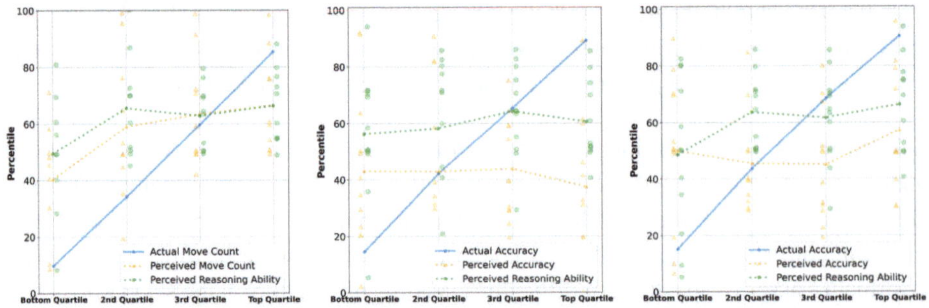

(a) Sliding Puzzle Game (b) Categorization Task: Car (c) Categorization Task: Credit

Fig. 5.19 DKE results from two studies across three tasks show actual vs. perceived performance percentiles on the X- and Y-axes, respectively. The blue line shows a baseline if participants perceived their performance accurately. The yellow dotted line depicts participants' perceived performance percentile in a sliding puzzle game (a) and a categorization task using car (b) and credit (c). The green color depicts participants' perceived reasoning ability. Across all three tasks, we observe the canonical curves indicative of DKE: bottom quartile performers tend to overestimate their performance, while top quartile performers tend to underestimate performance

count, further differentiated by task completion time as an additional performance indicator. Participants spent on average 3.18 minutes (min: 0.22, max: 27.23) to solve the puzzle.

H1: DKE in Visual Reasoning. To test **H1**, we assigned a percentile ranking for each participant based on their actual move count, then by time spent to complete the puzzle.

As Fig. 5.19a illustrates, bottom quartile participants whose actual move counts (blue) ranked in the 10th percentile on average placed themselves around the 40th percentile (yellow). In the top quartile, however, participants whose actual performance fell in the 85th percentile grossly underestimated their move count compared to their peers to be in the 65th percentile on average. We found a statistically significant discrepancy between the actual and perceived percentiles for both the bottom quartile ($t = -4.11$, $p < 0.01$) and the top quartile ($t = 4.62$, $p < 0.01$). The misjudgment of performance among the two extreme quartiles was still observed even though perceived performance was significantly correlated with actual performance ($r(37) = 0.36$, $p = 0.025$). This indicates that although there is a general alignment between individuals' perceptions of their performance and their actual performance, those at the very high and very low ends of the spectrum tend to have distorted views of their own performance. Thus, we find support for **H1**, consistent with DKE.

Discussion of Results. Our results indicated that DKE is observable in the context of visual reasoning in the sliding puzzle game. We also considered possible confounds, e.g., that participants who performed the fewest movements might not spend the least time completing the puzzle, as they may be more likely to spend more time strategizing before moving. However, further exploratory analysis suggests otherwise. On average, the top performers took 0.47 minutes to "think" before making their first move (the time between loading the puzzle game page and initiating the first move), compared to 1.57 minutes for the bottom group. In addition, the top performers averaged only 0.29 minutes ($SD = 0.14$) elapsed between the first and last move, whereas the bottom group took much longer with 6.65 minutes ($SD = 6.41$).

We also considered **time spent** as a measure of success (*"how quickly (less time is better) do you think you did relative to your peers"*), rather than move count, and nonetheless observed a similar pattern of DKE (see details in the supplemental materials). While the extent of the bias varied slightly between the two measures of success (move count vs. time spent), the overarching trend was consistent: participants at both ends of the skill spectrum showed discrepancies between their perceived and actual performances. Likewise, we also considered how performance compares to individuals' general perceived **reasoning ability** (*"how well do you think your reasoning ability compares to your peers"*) (Fig. 5.19a, green). We again found a comparable result among top ($t = 4.54$, $p < 0.01$) and bottom ($t = -5.15$, $p < 0.01$) quartile. This reinforces the prevalence of DKE across multiple measures of task success and general reasoning ability. Details are provided in the supplemental materials.

H2: Performance and Interactions. To test **H2**, we visually examined the interactions of participants in four quartiles using lines overlaid on the puzzle grid (Fig. 5.20). Line thickness is proportional to the number of times a tile was moved in the given direction. For example, a horizontal mark in the top right portion of the figure signifies that participants moved tiles left-right or right-left into the top rightmost grid cell of the puzzle. The thicker the line, the more frequently that path was taken. The line widths are normalized based on each participant's actual move counts to ensure the view of the interactions is not dominated by participants who performed significantly more moves than others in their quartile.

Because 23 participants achieved the optimal solution, Fig. 5.20 reveals the same movement paths for the third ($n_{Q3} = 10$) and top ($n_{Q4} = 9$) quartile groups, reflecting a single unique optimal solution. Low-skilled participants (Q_1) tended to randomly explore the board to find a solution, compared to their higher skilled peers. We thus find support for **H2** via visual inspection that there are detectable differences in interactive strategies used by individuals who are more and less competent.

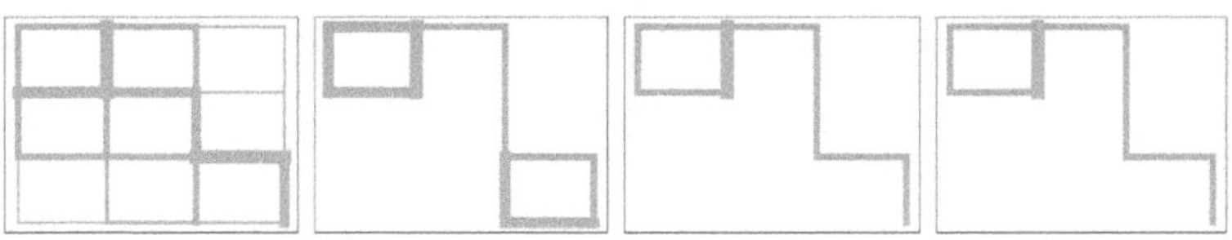

(a) Q_1 (b) Q_2 (c) Q_3 (d) Q_4

Fig. 5.20 Interaction strategies by four participant groups (Q_1 = lowest performers and Q_4 = highest performers)

Discussion of Results. The interaction analysis highlights differences in the interactions of participants across varying skill levels during the puzzle task. However, the ability to discern differences in strategies can be complicated by the large number of participants who achieved an optimal solution. Had we chosen a more complex puzzle configuration or one that had multiple optimal solutions, we may have been able to observe greater diversity in strategies among the top performers. We explore this further in the more difficult scatterplot categorization task in Study 2 (Sect. 5.6.3).

H3: Personality and Interactions. To test **H3**, we visualized participants' movement patterns similar to the analysis for **H2**, but stratified by low and high scores for personality traits rather than low and high task performance. We analyze each individual personality trait independently, consistent with prior work on personality traits and DKE [114] and personality traits and interactions [16]. We confirmed that the distribution of scores for all five personality traits were normally distributed using a Shapiro–Wilk test [117] with all p values > 0.05. According to [63], participants' personality trait scores are considered "average" if they fall within one-half standard deviation of the mean. Accordingly, we categorize the middle 40% of scores as average, with each tail (30%) representing high and low values for each personality trait. Figure 5.21 illustrates movement paths with average move counts by varied personality groups.

By visual inspection, we observe that participants with higher scores for each of four personality traits (conscientiousness, extraversion, agreeableness, and neuroticism) tended to explore the entire puzzle grid more evenly to find a solution and generally with higher move counts, whereas those with a lower score often left blank areas in certain segments of the grid. However, a significant difference was observed in move count between individuals with high and low scores for conscientiousness only ($u = 105$, $p = 0.01$).

These preliminary findings could indicate that individuals with lower scores might lean toward more precise interaction strategies and potentially achieve optimal task performance, although these trends are not statistically significant for four traits. Overall, we find mixed

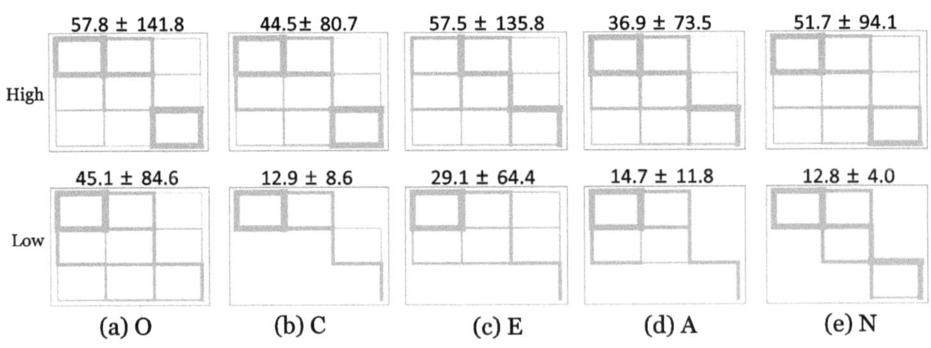

Fig. 5.21 Movement path triggered by different personality traits with move counts Mean ± SD

support for **H3**, with observable differences in interaction strategies for agreeableness, conscientiousness, extraversion, and neuroticism, and less clear differences for openness.

Discussion of Results. Our results suggest that participants scoring high in conscientiousness, extraversion, agreeableness, and neuroticism appeared to evenly explore the puzzle grid, possibly indicating a more exhaustive or trial-and-error approach to problem-solving. In contrast, participants with lower scores in these traits exhibited more selective interaction with the grid. One potential explanation for the blank areas in their movement paths could indicate a more contemplative approach where participants might have spent time strategizing before making moves which could reflect a more cautious or measured approach to problem-solving. The mixed support for **H3** underscores the complexity of the relationship between personality and problem-solving behaviors and emphasizes the need for further research to unpack these dynamics.

H4: Personality and Δ Performance. To test **H4**, we computed Pearson correlation [104] for each of five personality traits compared to their respective disparity between actual and perceived performance. We use difference in performance as a proxy for susceptibility to DKE, e.g., people with a larger magnitude of Δ Performance are more prone to metacognitive miscalibration. Figure 5.22 depicts personality trait scores (x-axis, ranging from 4 to 20), and Δ Performance (y-axis, $\Delta =$ Estimated Percentile $- -$Actual Percentile). When there is a positive/negative slope, it suggests a trend in over-/ underestimation of performance relative to the personality trait, while a flatter slope reflects more accurate perception of performance independent of personality traits.

Only Conscientiousness (C) was observed to have a significant effect on perceived performance ($r(37) = 0.48$, $p < 0.01$), implying that individuals with higher conscientiousness scores tend to exhibit a greater magnitude of overestimation of task performance. Thus, we find weak support for **H4**.

Discussion of Results. We observe that individuals high in conscientiousness, known for their meticulousness and strong commitment to task completion, may exhibit an optimistic attitude in assessing their capabilities and achievements [111, 124]. This optimism could stem from lofty personal standards and goals, leading to a self-view that matches their ideal performance.

While we found a statistically significant correlation for conscientiousness, we note that this result may be sensitive to how we define top and bottom performers. For example, altering our selection method to a random choice of 9 from the 23 participants who attained the optimal move count as the top quartile—instead of further differentiating by task completion time—eliminated the statistical significance for all personality traits (see details in Supplemental Materials). This suggests that the significant association with conscientiousness may not be robust and could be dependent on the performance metrics we adopted.

H5: Δ Performance and Domain. To test **H5**, we investigated the correlation between self-reported domain familiarity (from 1 to 5) and the manifestation of DKE. The Pearson corre-

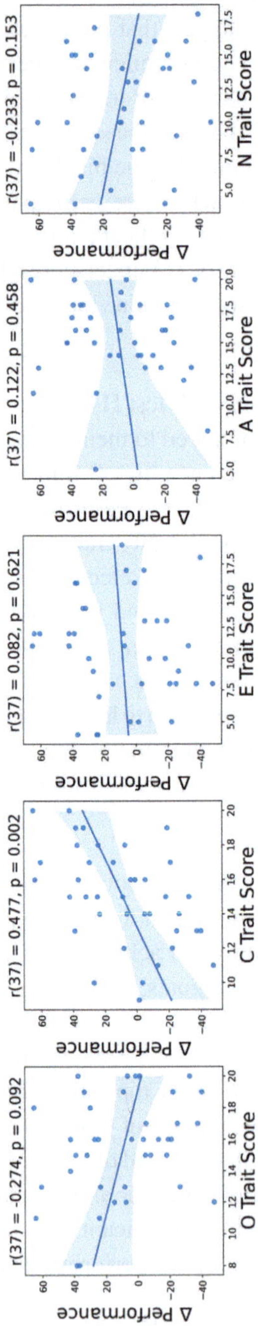

Fig. 5.22 Correlation between personality traits and ΔPerformance = Estimated Percentile − −Actual Percentile)

lation revealed no significant correlation between the two variables ($r(37) = -0.035$, $p = 0.83$). Thus, we find no support for **H5**.

Discussion of Results. Our results suggest that the tendency for people to overestimate their performance is not correlated with their familiarity with the sliding puzzle game. These results contribute to the ongoing discourse about the complex nature of self-assessment in cognitive tasks and highlight the interplay between self-perception and actual skill levels in various domains [27, 52].

5.6.3 Study 2: Categorization with Interactive Scatterplot

In Study 1, we replicated DKE in the context of a puzzle game, in which bias was measured as a function of perceived optimality of achieving the solution. We explore a complementary task in Study 2, where participants categorize data points in an interactive scatterplot. Bias in this study is measured as a function of the perceived accuracy of categorization. We build upon Study 1 by seeking to replicate DKE in a task that increases the complexity of interactions supported and gets closer to a realistic judgment task facilitated by visualizations. We present results of a pre-registered within-subject study exploring DKE in the context of two categorization tasks in the domains of cars and credit where participants engage in interactive labeling [78].

5.6.3.1 Experimental Setup

Dataset and Tasks. Participants completed two tasks (order counterbalanced) in different domains. We used datasets from the domains of car type and credit score level as the general public usually has a reasonable degree of familiarity with these topics. For the car task, participants were asked to assign one of three types (SUV, Sedan, Minivan) to each point by comparing the provided statistics for each car, such as engine size and fuel economy (see details in the supplemental materials). Similarly, for the credit task, participants were asked to assign one of the three levels (Good, Standard, Poor) to each point based on credit-related traits for each person, such as number of credit cards and credit history age (see details in the Supplemental Materials).

These domains allow us to understand (1) the generalizability of DKE in this task beyond a single domain and (2) varying levels of task complexity, facilitating the investigation of how these differences influence task performance [19].

To guarantee that the classes are distinctly separable based on the attributes of the data, we selected 30 points from each dataset and selected a subset of attributes to describe each data point: 6 attributes for the car task and 8 attributes for the credit task. This served as a proxy for task difficulty, where a pilot study with n = 12 participants confirmed (t =

4.69, $p < 0.01$) that the credit task ($\mu_{accuracy} = 0.33$) was more difficult than the car task ($\mu_{accuracy} = 0.47$).

Interface. We used an interactive scatterplot system in which the primary view displays the 30 points that represent individual cars or bank customers (Fig. 5.23a). Hovering on a point shows details about the particular car or customer in a tooltip (B). To label a point in the scatterplot, participants can click the appropriate category button (C) then click the respective points in the scatterplot. The x- and y-axes can be changed to represent any of the attributes through a drop down menu (D). Task instructions and interface guidance are presented in a tooltip when participants hover over the help button (E). As with the 15-puzzle game in Study 1, interactions with the system were logged including time stamped records of click, hover, and axis interactions. To ensure data quality, we required participants to classify at least 90% of data points ($\geq$ 27) before proceeding to the next task.

Procedure. Participants were first presented with a practice task using a dog breeds dataset (categorize the dogs by breed: Bernedoodle, Shih Tzu, and American Bulldog based on attributes such as amount of shedding and size) to become comfortable with the interface prior to completing the main tasks. Prior to beginning each task, participants needed to select the default attributes that would be displayed on the x- and y-axes for the initial visualization, to avoid biasing participants to use any particular attributes in their decision-making. Additionally, we prefaced the task with an additional visualization literacy assessment including 7 multiple-choice questions (raw scores ranged from 0 to 7) specific to scatterplots adopted from VLAT [88] in the preliminary survey to ensure that participants could accurately interpret the scatterplot visualization.

Recruitment. We initially recruited 48 participants through the Prolific crowdsourcing platform based on a power analysis of pilot data with 12 participants who completed the categorization task using a similar experimental setup. Our minimum target sample size was 44 participants to obtain .8 power to detect a medium effect size of .25 at the standard $\alpha = 0.05$. Participants were initially compensated \$3.50 for the study which had an estimated duration of 20 minutes (actual median completion time of 25 minutes led to an adjustment of payment to \$4.85). Prolific allowed participants to spend a maximum of 67 minutes on the task; three participants timed out and their data was subsequently excluded. Participants were incentivized with an additional \$1 performance bonus, respectively, if they (1) completed a visualization literacy assessment survey with the highest correctness in the top 5% and (2) completed the categorization task with accuracy in the top 5%. 13 participants earned performance bonuses: 7 for top performance on the visualization literacy assessment, 3 for the highest accuracy in the car task, and another 3 for the highest accuracy in the credit task.

No data were excluded from analyses due to failed attention checks in this study. However, we excluded data from participants with a visualization literacy score below 3 out of 7, deviating from our initial pre-registration plan of measuring DKE as a function of visualization literacy. This decision was based on the realization that a fundamental grasp of

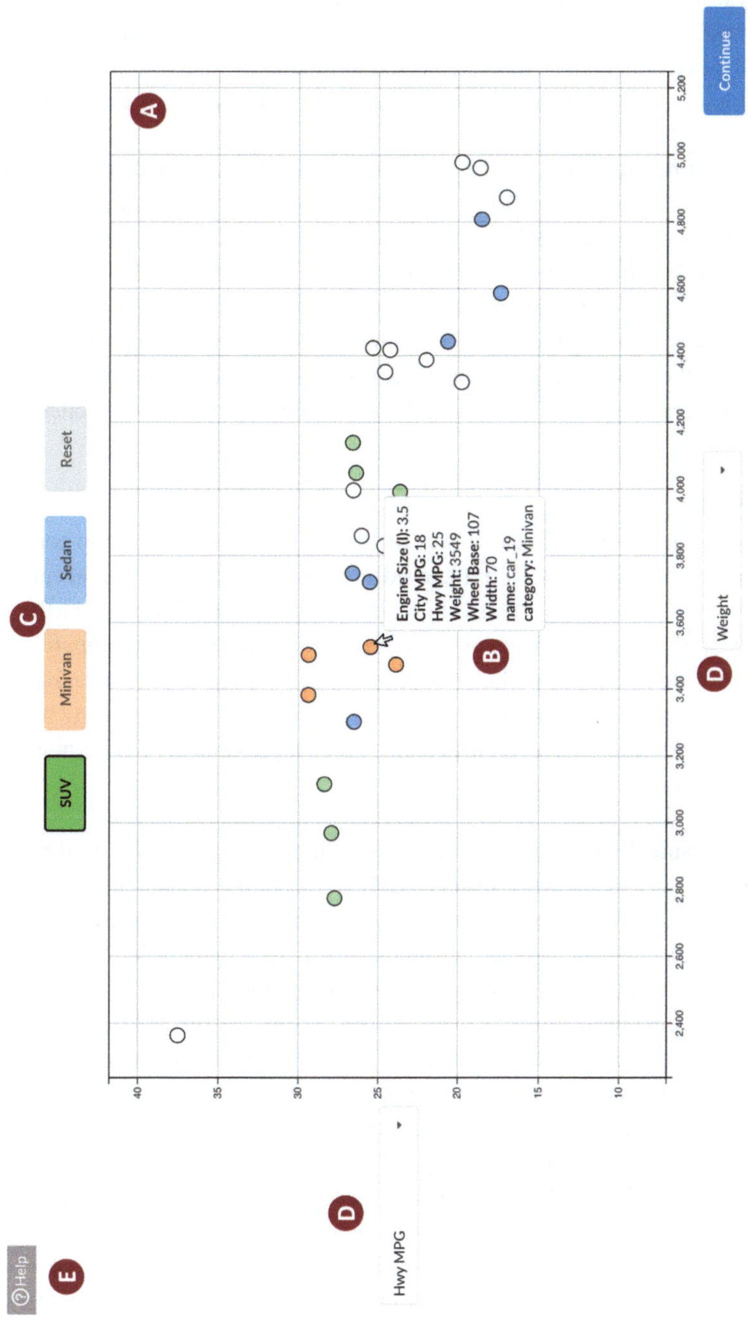

Fig. 5.23 Interactive scatterplot **a** that shows tooltips on hover **b**. It also features category buttons for labeling **c**, x- and y-axes dropdowns **d** and a help reminder of interface mechanics **e**

visualization literacy is a crucial prerequisite for meaningfully measuring DKE. Including participants with poor visualization literacy could undermine the study's integrity, akin to measuring DKE through a literature test presented in a language unfamiliar to the participants. Finally, two data points were excluded as outliers (outside 1.5 times the Interquartile Range). In total we used data from 46 individuals in our forthcoming analyses.

Participant Demographics. 12 participants identified as female, 33 identified as male, and 1 identified as non-binary. The majority of participants (31) hold a college degree (Bachelor's, Master's, or Doctorate), while 9 have some college or Associate's degree, and 6 have a High School diploma. Participants were on average 30.60 years old (SD = 9.48). After excluding individuals who achieved a scatterplot literacy score of less than 3 out of 7, the remaining participants achieved average scores of 5.18 (min = 3, max = 7) after applying the correction-for-guessing method [33, 53]. Participants reported average familiarity of 2.57 out of 5 for cars (35 participants rated 3 or lower) and 2.98 for credit (29 participants rated 3 or lower).

5.6.3.2 Results

In this study, we define performance based on categorization accuracy. Top performers (top quartile) are those individuals who achieved the highest categorization accuracy (largest number of points labeled correctly), while bottom performers (bottom quartile) are those who achieved the lowest categorization accuracy. To evaluate DKE, we compare participants' actual categorization accuracy to their perceived accuracy relative to their peers. Participants achieved on average 46.70% accuracy ($SD = 18.30\%$) in the car task and 32.59% accuracy ($SD = 10.20\%$) in the credit task. Participants spent approximately 3.47 minutes on the car task and 5.36 minutes on the credit task.

H1: DKE in Visual Reasoning. Consistent with the findings of DKE in the sliding puzzle game in Study 1, we likewise observe DKE for both the car and credit categorization tasks (Fig. 5.19(b and c, respectively)). Participants in the bottom quartile of the car task (Fig. 5.19b, blue), with an average accuracy in the 15th percentile, overestimated their accuracy (42th percentile, Fig. 5.19b, yellow) relative to their peers ($t = -3.36$, $p < 0.01$). Conversely, participants in the top quartile, who scored in the 90th percentile on average, significantly underestimated their accuracy (37th percentile) relative to their peers ($t = 6.19$, $p < 0.01$). A congruent pattern was observed in the credit task (Fig. 5.19c) as well, where bottom quartile participants, scoring in the 15th percentile (Fig. 5.19c), blue) on average, overestimated their accuracy (50th percentile, Fig. 5.19c, yellow) relative to their peers ($t = -4.27$, $p < 0.01$), while top quartile participants, scoring in the 90th percentile on average underestimated their accuracy (57th percentile) relative to their peers ($t = 4.95$, $p < 0.01$). This finding supports **H1**, less competent individuals overestimate their performance relative to peers, while competent individuals underestimate their performance.

Discussion of Results. These findings offer an empirical understanding of the relationship between self-assessment and actual performance in both low-skilled and high-skilled participants in an interactive scatterplot categorization task across two domains, contributing to the growing body of knowledge that DKE is a generalized pattern rather than a task-specific anomaly. Similar to Study 1, we considered an individual's general perceived **reasoning ability** relative to their peers (Fig. 5.19(b and c, green)) and again found comparable results (car: bottom quartile ($t = -6.11$, $p < 0.01$), top quartile ($t = 5.45$, $p < 0.01$); credit: bottom quartile ($t = -4.00$, $p < 0.01$), top quartile ($t = 4.55$, $p < 0.01$)).

We note that accuracy in the credit task was chance, suggesting higher task difficulty. This could be due in part to the greater number of attributes increasing task complexity. It is unlikely participants categorically misunderstood the task given (1) the higher accuracy (47%) in the car task and (2) the higher average self-reported credit familiarity level ($\mu_{car} = 2.57$ vs. $\mu_{credit} = 2.98$) with 22 participants rating 3 or above for cars and 32 participants rating 3 or above for credit. Detailed distributions can be found in supplemental materials.

H2: Performance and Interactions. To test **H2**, our analysis was divided into three components of interactive behavior, including (1) rate of interaction with data labeling; (2) think time preceding an interaction; and (3) interaction sequences, which can reflect potential strategies employed by participants.

Interaction Rate. Previous research has identified distinct patterns in user behavior where some individuals tend to engage in a meticulous contemplation of each action, resulting in slow, deliberative manipulation of input devices, whereas others opt for more rapid execution [62]. In this study, top and bottom quartile participants in the car task registered 8.75 and 7.55 interactions per second, respectively ($t = -0.566$, $p = 0.58$), while in the credit task, the corresponding rates were 6.02 and 8.19 interactions per second ($t = -0.541$, $p = 0.31$). In both scenarios, no significant differences in interaction rate were detected among top and bottom quartile.

Think Time. The *think time* metric is operationalized as the temporal interval from the end of one interaction to the initiation of the subsequent interaction. We found a significant difference in think time between two consecutive interactions for the top and bottom quartiles only in the credit task ($u = 1826896$, $p < 0.01$) using Mann–Whitney U test [91]. We further deconstructed this analysis by interaction type (e.g., considering clicks and hovers separately). As depicted in Fig. 5.24, the left segment of each figure illustrates the mean think time preceding each type of interaction [6], and the right segment enumerates the average counts of the corresponding interaction types [6] for the car (a) and credit (b) tasks. Given that multiple pairwise comparisons were made, the Bonferroni correction [40] was applied to control the family-wise error rate, adjusting the significance level from 0.05 to 0.01. For think time by interaction type, no significant differences were observed between the two quartiles. Similarly, no significant differences in interaction counts were found between the two groups.

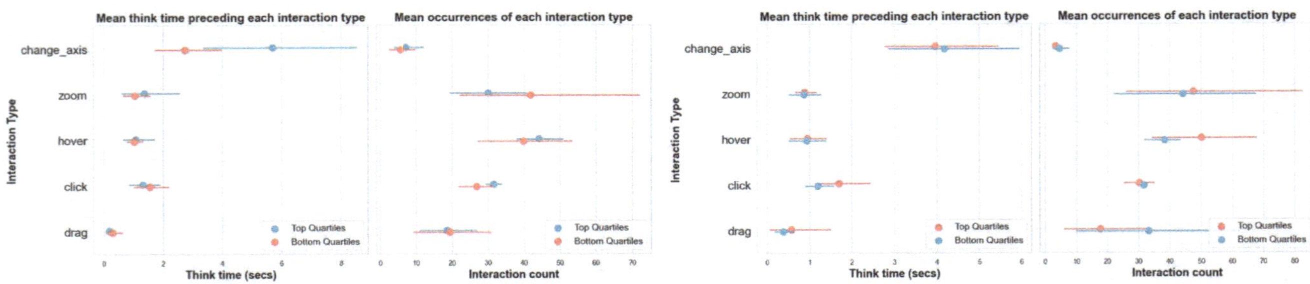

(a) Car (b) Credit

Fig. 5.24 Average think time (left) preceding interaction types with 95% confidence interval and corresponding average interaction counts (right) with 95% confidence interval for the car (a) and credit (b) tasks

Interaction Sequences. Interaction *sequences* could reveal underlying strategies employed by users during the categorization task. To assess this, we computed a transition matrix, aimed at uncovering common sequential patterns of interactions. Figure 5.25 demonstrates the frequency of one interaction type (x-axis) succeeded by another interaction type (y-axis), stratified by both high-skilled participants and low-skilled participants in car (a) and credit (b) tasks. Each element within the transition matrix has been normalized relative to the cumulative sum of its corresponding row. This normalization means that each element now represents a proportion of the total for that row, ensuring a fair comparison of transition probabilities between different types of interactions (i.e., significantly more "hover" interactions occurred compared to "change axis" interactions). Cells within the matrix are color coded, with darker shades signifying sequences of interaction types that occur with greater frequency. We then flattened each matrix into one-dimensional vectors and calculated the Pearson correlation coefficient [104] between the two vectors, yielding a value of $r(22) = 0.93$, $p < 0.01$ for the car task, $r(22) = 0.91$, $p < 0.01$ for the credit task. The significant correlation between the two vectors suggests there were no real differences in interaction patterns across bottom and top quartiles in both tasks.

Discussion of Results. While discernible differences appeared in interaction patterns in the 15-puzzle game in Study 1, we observe relatively little distinction between interactive strategies of top and bottom quartile participants in the categorization tasks. We observed no differences in interaction rate or interaction sequences, and marginal differences in think time for some specific interaction types. Collectively, these findings provide little support for **H2** that there are detectable differences in interactive strategies used by individuals who are more and less competent. Inconsistent differences observed across some measures of interactive strategy may be attributable to the heightened complexity of the task and breadth of available interactions leading to individual differences in task approach.

H3: Interaction and Personality. To test **H3**, we performed similar analysis as **H2**, focusing on three key dimensions of participants' decision-making behavior: interaction rate, think time, and interaction sequences, but in this analysis stratified by low- and high-score categories for five personality traits. Similar to Study 1, participants' personality trait scores were standardized.

Interaction Rate. We computed the interaction rate for high and low scores on five personality traits. The results are shown in Table 5.8. We applied Bonferroni correction [40] to control the family-wise error rate, adjusting the significance level from 0.05 to 0.01. However, no significant differences in interaction rates were observed between high and low scores for the five personality traits in either task.

Think Time. In an effort to understand the determinants of think time preceding each type of interaction, we fit linear mixed-effects models, with interaction type, personality traits, and their interaction terms as fixed effects, and participants as a random effect. Five categories

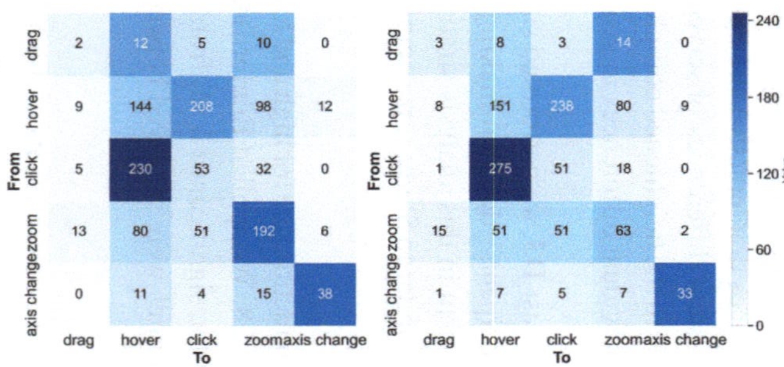

(a) Car Task: bottom (left) and top (right) quartiles

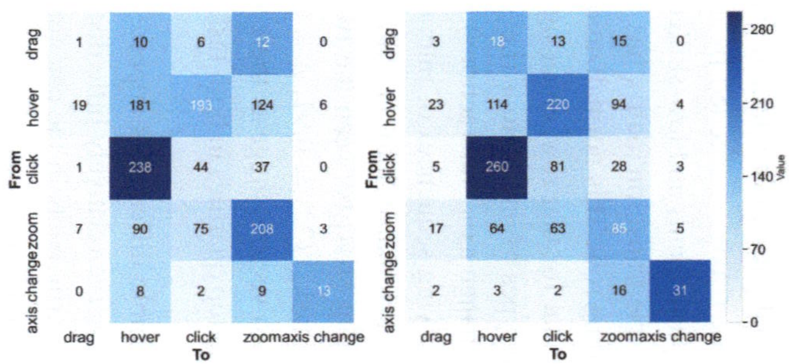

(b) Credit Task: bottom (left) and top (right) quartiles

Fig. 5.25 Transition matrix representing interaction sequences

Table 5.8 Interaction rate (interactions per second) for individuals with high/low scores on five personality traits

	O		C		E		A		N	
Score/Task	Car	Credit	Car	Credit	Car	Credit	Car	Credit	Car	Credit
High	9.20	7.22	10.19	7.23	8.54	7.38	8.25	6.68	9.07	9.25
Low	9.43	6.14	8.61	8.88	10.21	6.83	10.67	8.70	8.81	7.37

Table 5.9 Car: Determinants of think time

	O		C		E		A		N	
Task/Score	High	Low	High	Low	High	Low	High	Low	High	Low
Change axis	–0.31	–1.40	**1.78***	**2.04****	1.08	**2.45***	–0.66	0.77	1.59	–0.43
Drag	0.81	0.52	–0.10	0.3	0.33	0.53	–0.39	–0.30	–0.43	0.38
hover	0.21	**0.92***	–0.17	0.56	0.37	**1.09***	0.19	–0.15	**–0.97***	0.03
Zoom	–0.32	0.12	–0.11	0.83	–0.19	0.66	0.72	0.25	–0.92	0.20

Table 5.10 Credit: Determinants of think time

	O		C		E		A		N	
Task/Score	High	Low	High	Low	High	Low	High	Low	High	Low
Change axis	0.56	**3.13****	–0.91	–0.08	–0.98	**3.18****	–1.55	**–3.32****	–0.34	–0.17
Drag	–0.08	–0.01	–0.11	1.60	–0.28	0.32	–0.13	0.16	**–1.90****	0.09
hover	0.54	0.44	**0.74***	0.36	–0.62	–0.34	–0.59	–0.19	–0.44	–0.25
Zoom	0.09	0.15	0.46	**–1.56****	**–0.78***	–0.17	0.03	–0.47	–1.01	0.22

of interaction types ("click," "drag," "hover," "zoom," and "change axis") were included in the model as a categorical variable and the interaction type "click" was used as the reference category in the coding scheme.

We can see the results stratified by high and low scores in the five personality traits for the car task (Table 5.9) and credit task (Table 5.10). A lower score in Extraversion exhibits a significantly positive impact on the think time preceding the "change axis" interaction in both tasks. Other traits also have significant effects on think time for one of the tasks. For example, in the car task, both higher and lower scores in the Conscientiousness trait are significantly positively correlated with think time preceding change axis interaction. This could suggest that the effect of Conscientious on think time is significantly modulated by the type of interaction. Specifically, the "change axis" interaction might require more deliberation, making typically less conscientious individuals spend more time thinking, which highly conscientious individuals are inclined toward. However, these effects vary across the two tasks.

Interaction Sequence. We examined the interaction sequence and interaction attention among participants scoring high/low on five personality traits. We computed the transition matrix to identify underlying sequential patterns exhibited by the high- and low-scoring individuals. Significant differences were not detected in either of the two tasks.

Discussion of Results. Although significant effects were observed for think time for specific personality traits combined with certain interaction types, these effects varied between the two tasks. Despite sharing commonalities in nature and setting, the unique domain-specific attributes and possible differences in task complexity likely contributed to these divergent outcomes. Thus, it is crucial to recognize that while some findings are statistically significant, they are specific to the contexts we studied. Overall, these findings provide mixed support for **H3**.

H4: Personality and Δ Performance. To test **H4**, we conducted a Pearson correlation analysis to discern if there was a correlation between individual personality traits and the susceptibility to DKE (Δ Performance = Estimated Percentile–Actual Percentile). As depicted in Fig. 5.26, only the Conscientiousness (C) trait showed a significant effect in the car task ($r(44) = 0.431$, $p < 0.01$). While we observe some nonzero trend lines for other traits, the differences are not statistically significant. These findings provide little support for **H4**, that there are some correlations between personality traits and task performance.

Discussion of Results. While we observed a notable relationship between Conscientiousness and miscalibration of perceived performance, this pattern was not evident in the credit task, which may suggest a context-dependent relationship, e.g., confounded by task difficulty.

H5: Δ Performance and Domain. To test **H5**, we consolidated the analysis across both the car and credit tasks to probe for a correlation between self-reported domain familiarity and the manifestation of DKE.

The Pearson correlation analysis revealed a significant positive correlation between the difference in performance and domain familiarity ($r(90) = 0.448$, $p < 0.01$) (see supplemental materials). This finding suggests that individuals who perceive themselves as more familiar with a specific domain may be likely to overestimate their abilities within that domain, whereas those who report less familiarity may conversely underestimate their capabilities, supporting **H5**, that people's overestimation of their performance is positively associated with their familiarity of the domain.

Discussion of Results. Individuals with higher self-reported familiarity in a specific domain were found to overestimate their abilities, which, in a broad sense, aligns with the general trend of DKE that people with lower ability at a task (using domain familiarity as a proxy for ability) tend to overestimate their ability [41]. On the other hand, the lack of significant correlation in Study 1 suggests that this trend might not be universally applicable across all

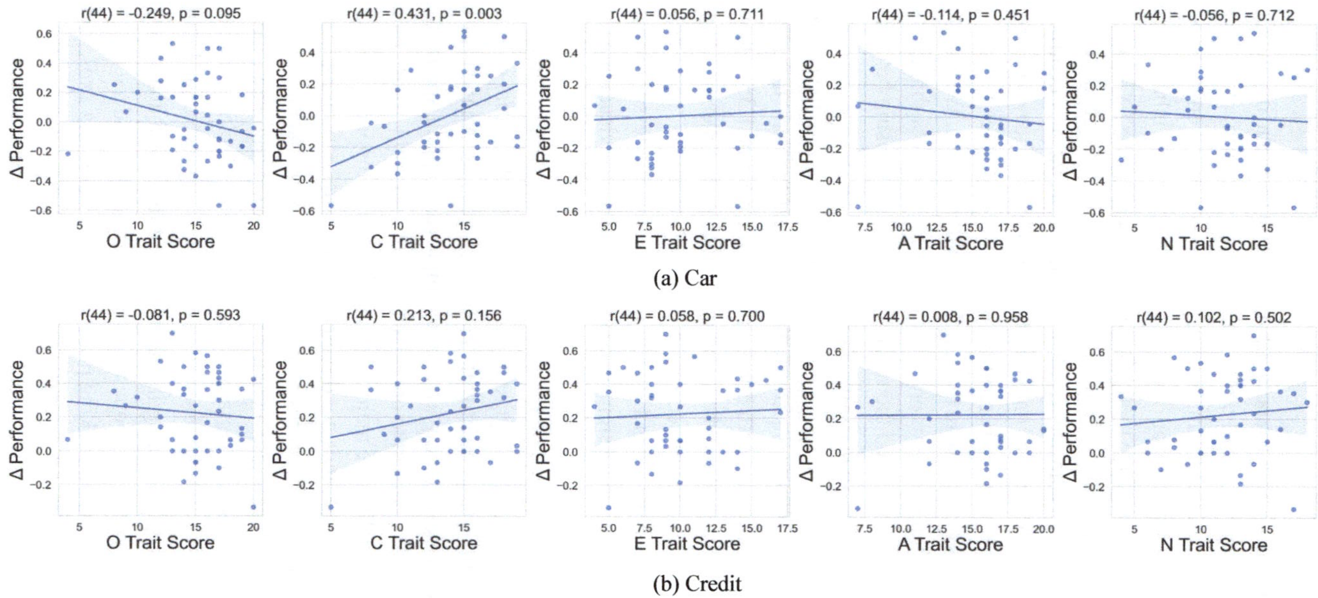

Fig. 5.26 Correlation between personality traits and ΔPerformance $=$ Estimated Percentile $- -$Actual Percentile for **a** the car task and **b** the credit task

domains or that other factors may influence the relationship between domain familiarity and accuracy of self-assessment.

Exploratory Analysis. To understand the extent to which our findings are attributable to the use of visualization specifically, rather than simply an artifact of the labeling task, we conducted additional exploratory analyses to analyze how users engaged with the axes of the scatterplot by selecting different attribute pairs. Some pairs are inherently more informative for the tasks, i.e., would produce more clear clustering of correctly labeled points. To quantify this, we calculated the ratio of inter-class to intra-class distances for each attribute pair. Here, inter-class distance is the average distance between category centroids, and intra-class distance measures the distance of data points from their category's centroid [94]. A higher ratio signifies better category separability, which we will refer to as a more informative attribute pair. Analysis details are in the supplemental materials. We found that top performers tended to more often choose more informative attribute combinations, suggesting that the interactions with the axes and the resulting visual relationship between attributes likely contributed to the disparity in performance.

For instance, in the car task, top performers most often chose the combination Weight × Engine Size with a ratio of 2.6 compared to the bottom performers' most chosen combination of Wheel Base × Engine Size with a less informative ratio of 1.35. Likewise, in the credit task, top performers most often chose the combination Number of Loans × Outstanding Debt with a ratio of 2.54 compared to the bottom performers' most chosen combination of Monthly Balance × Number of Delayed Payments with a less informative ratio of 1.02. These disparities in ratios could signify the successful use of visual clustering strategies for the labeling task.

Our findings further showed that in both tasks, 30 out of 46 participants interacted with the axes more than the default requirement (twice, to set the initial configuration), indicating a more meaningful engagement with the interactive axes. Analysis of these participants revealed varied strategies: some participants employed explicit spatial techniques, as evidenced by feedback like, "I tried to organize them into sensible groups using the axis to sort them," highlighting a deliberate manipulation of visual components. In contrast, others followed more ambiguous methods not directly tied to visualization, e.g., one participant said "I decided to categorize credit scores based on their monthly balance and number of delayed payments, I believe. I think these were the most important among all other factors." While we cannot assert with absolute certainty that mentioning specific attribute pairs equates to direct interaction with visual elements, such references suggest an inclination toward visual analysis rather than simply labeling in the absence of the visual analysis setting. Moreover, the widespread use of visual components among participants underscores the critical role of visualization in their decision-making process. By manipulating axes, identifying patterns, and grouping data visually, users are engaging in a form of visual reasoning that leverages spatial relationships and graphical representations to draw conclusions.

We also found that DKE persisted even when we focused our exploration on the subset of 30 participants who interacted more heavily with the axes, doing so at least twice. Particularly, we observed a significant overestimation of performance by the lower performing group in both the car task ($t = -2.15$, $p = 0.049$) and the credit task ($t = -3.09$, $p < 0.01$). Conversely, the higher performing group significantly underestimated their abilities in both the car ($t = 4.55$, $p < 0.01$) and credit ($t = 3.07$, $p < 0.01$) tasks. Collectively, these exploratory findings increase our confidence that the observed phenomenon is attributable to DKE in judgments facilitated by visualization.

References

1. Robert Amar, James Eagan, and John Stasko. "Low-level components of analytic activity in information visualization". In: *IEEE Symposium on Information Visualization, 2005. INFOVIS 2005.* IEEE. 2005, pp. 111–117.
2. John Archer. "Why do people love their pets?" In: *Evolution and Human behavior* 18.4 (1997), pp. 237–259.
3. Karthik Badam et al. "Supporting Team-First Visual Analytics through Group Activity Representations". In: *Graphics Interface* (2017).
4. Mahzarin R Banaji and Anthony G Greenwald. "Implicit gender stereotyping in judgments of fame." In: *Journal of personality and social psychology* 68.2 (1995), p. 181.
5. Scott Bateman et al. "Useful junk? The effects of visual embellishment on comprehension and memorability of charts". In: *Proceedings of the SIGCHI conference on human factors in computing systems.* 2010, pp. 2573–2582.
6. Leilani Battle and Jeffrey Heer. "Characterizing exploratory visual analysis: A literature review and evaluation of analytic provenance in tableau". In: *Computer graphics forum.* Vol. 38. 3. Wiley Online Library. 2019, pp. 145–159. DOI: https://doi.org/10.1111/cgf.13678.
7. Cindy Xiong Bearfield et al. "Same Data, Diverging Perspectives: The Power of Visualizations to Elicit Competing Interpretations". In: *IEEE Transactions on Visualization and Computer Graphics* (2024).
8. Elaine Blakey and Sheila Spence. *Developing metacognition.* ERIC Clearinghouse on Information and Technology, 1990.
9. Corwin Boake. "From the Binet–Simon to the Wechsler–Bellevue: Tracing the history of intelligence testing". In: *Journal of clinical and experimental neuropsychology* 24.3 (2002), pp. 383–405.
10. Brett J Borghetti and Christina F Rusnock. "Introduction to Real-Time State Assessment". In: *Foundations of Augmented Cognition: Neuroergonomics and Operational Neuroscience.* Vol. 9743. Lecture Notes in Computer Science. Springer, 2016, pp. 311–321.
11. Michelle J Bovin et al. "Psychometric properties of the PTSD checklist for diagnostic and statistical manual of mental disorders–fifth edition (PCL-5) in veterans." In: *Psychological assessment* 28.11 (2016), p. 1379.
12. Jeremy Boy et al. "A principled way of assessing visualization literacy". In: *IEEE transactions on visualization and computer graphics* 20.12 (2014), pp. 1963–1972.
13. Gregory J Boyle. "Myers-Briggs type indicator (MBTI): some psychometric limitations". In: *Australian Psychologist* 30.1 (1995), pp. 71–74.
14. Laura A Brannon, Michael J Tagler, and Alice H Eagly. "The moderating role of attitude strength in selective exposure to information". In: *Journal of Experimental Social Psychology* 43.4 (2007), pp. 611–617.

15. Amanda B Breen and Andrew Karpinski. "Implicit and explicit attitudes toward gay males and lesbians among heterosexual males and females". In: *The Journal of social psychology* 153.3 (2013), pp. 351–374.
16. Eli T Brown et al. "Finding waldo: Learning about users from their interactions". In: *IEEE Transactions on visualization and computer graphics* 20.12 (2014), pp. 1663–1672. DOI: https://doi.org/10.1109/TVCG.2014.2346575.
17. Eli T. Brown et al. "Dis-Function: Learning Distance Functions Interactively". In: *IEEE Conference on Visual Analytics Science and Technology (VAST)* (2012), pp. 83–92.
18. Adrian Brüngger et al. "The parallel search bench ZRAM and its applications". In: *Annals of Operations Research* 90.0 (1999), pp. 45–63. DOI: https://doi.org/10.1023/A:1018972901171.
19. Katherine A Burson, Richard P Larrick, and Joshua Klayman. "Skilled or unskilled, but still unaware of it: how perceptions of difficulty drive miscalibration in relative comparisons." In: *Journal of personality and social psychology* 90.1 (2006), p. 60. DOI: https://doi.org/10.1037/0022-3514.90.1.60.
20. Ruth MJ Byrne. "Counterfactual thought". In: *Annual review of psychology* 67.1 (2016), pp. 135–157.
21. Quinn Capers IV et al. "Implicit racial bias in medical school admissions". In: *Academic Medicine* 92.3 (2017), pp. 365–369.
22. Steve D Charman, Melissa Kavetski, and Dana Hirn Mueller. "Cognitive bias in the legal system: Police officers evaluate ambiguous evidence in a belief-consistent manner". In: *Journal of Applied Research in Memory and Cognition* 6.2 (2017), pp. 193–202.
23. Muhammad Faisal Cheema, Stefan Jänicke, and Gerik Scheuermann. "AnnotateVis: Combining traditional close reading with visual text analysis". In: *Workshop on Visualization for the Digital Humanities, IEEE VIS*. 2016.
24. Yu-Chi Chen, Chi-Yu Chao, and Huei-Tse Hou. "Learning pattern recognition skills from games: Design of an online pattern recognition educational mobile game integrating algebraic reasoning scaffolding". In: *Journal of Educational Computing Research* 61.6 (2023), pp. 1232–1251. DOI: https://doi.org/10.1177/07356331231171622.
25. Mengyu Chen, Yijun Liu, and Emily Wall. "Unmasking Dunning-Kruger Effect in Visual Reasoning and Visual Data Analysis". In: *IEEE Transactions on Visualization and Computer Graphics* (2024). to appear.
26. Mengyu Chen and Emily Wall. "Perception of Skill in Visual Problem Solving: An Analysis of Interactive Behaviors, Personality Traits, and the Dunning-Kruger Effect". In: *TREX: Workshop on TRust and EXpertise in Visualization* (2022).
27. Leonore Hildegonde Christoph et al. *The role of metacognitive skills in learning to solve problems*. SIKS, 2006.
28. Moreno I Coco and Rick Dale. "Cross-recurrence quantification analysis of categorical and continuous time series: An R package". In: *Frontiers in Psychology* 5 (2014), p. 510.
29. Maia B Cook and Harvey S Smallman. "Human factors of the confirmation bias in intelligence analysis: Decision support from graphical evidence landscapes". In: *Human Factors* 50.5 (2008), pp. 745–754.
30. Joseph A. Cottam and Leslie M. Blaha. "Bias by default? A means for a priori interface measurement". In: *DECISIVe: Workshop on Dealing with Cognitive Biases in Visualizations* (2017).
31. Marcel Danesi and Marcel Danesi. "Puzzles and Spatial Reasoning". In: *Ahmes' Legacy: Puzzles and the Mathematical Mind* (2018), pp. 105–125. DOI: https://doi.org/10.1007/978-3-319-93254-5_4
32. Ariel Deardorff. "Tableau (version. 9.1)". In: *Journal of the Medical Library Association* 104.2 (2016), pp. 182–183.
33. James Diamond and William Evans. "The correction for guessing". In: *Review of educational research* 43.2 (1973), pp. 181–191. DOI: https://doi.org/10.3102/00346543043002181.

34. David L Dickinson. "Political ideology, mood response, and the confirmation bias". In: (2022).
35. Evanthia Dimara et al. "A task-based taxonomy of cognitive biases for information visualization". In: *IEEE transactions on visualization and computer graphics* (2018).
36. Evanthia Dimara et al. "Mitigating the attraction effect with visualizations". In: *IEEE transactions on visualization and computer graphics* 25.1 (2019), pp. 850–860.
37. Frank Dobbin and Alexandra Kalev. "Why doesn't diversity training work? The challenge for industry and academia". In: *Anthropology Now* 10.2 (2018), pp. 48–55.
38. Helmut Doleisch et al. "Interactive focus+ context analysis of large, time-dependent flow simulation data". In: *Simulation* 82.12 (2006), pp. 851–865.
39. M Brent Donnellan et al. "The mini-IPIP scales: tiny-yet-effective measures of the Big Five factors of personality." In: *Psychological assessment* 18.2 (2006), p. 192.
40. Olive Jean Dunn. "Multiple comparisons among means". In: *Journal of the American statistical association* 56.293 (1961), pp. 52–64. DOI: https://doi.org/10.1080/01621459.1961.10482090.
41. Joyce Ehrlinger et al. "Why the unskilled are unaware: Further explorations of (absent) self-insight among the incompetent". In: *Organizational behavior and human decision processes* 105.1 (2008), pp. 98–121. DOI: https://doi.org/10.1016/j.obhdp.2007.05.002.
42. Madison A Elliott et al. "A design space of vision science methods for visualization research". In: *IEEE Transactions on Visualization and Computer Graphics* 27.2 (2020), pp. 1117–1127.
43. A Endert et al. "The State of the Art in Integrating Machine Learning into Visual Analytics". In: *Computer Graphics Forum*. Wiley Online Library. 2017.
44. Alex Endert et al. "The human is the loop: new directions for visual analytics". In: *Journal of Intelligent Information Systems* 43.3 (2014), pp. 411–435.
45. D. C. Engelbart. *Augmenting Human Intellect: A Conceptual Framework*. Tech. rep. AFOSR-3223. Menlo Park, CA: Stanford Research Institute, 1962.
46. Mussweiler Englich and T Mussweiler. "Anchoring effect". In: *Cognitive Illusions: Intriguing Phenomena in Judgement, Thinking, and Memory* (2016), p. 223.
47. Franz Faul et al. "G* Power 3: A flexible statistical power analysis program for the social, behavioral, and biomedical sciences". In: *Behavior research methods* 39.2 (2007), pp. 175–191.
48. Mi Feng et al. "Hindsight: Encouraging exploration through direct encoding of personal interaction history". In: *IEEE transactions on visualization and computer graphics* 23.1 (2016), pp. 351–360.
49. Alberto Ferrari and Marco Russo. *Introducing Microsoft Power BI*. Microsoft Press, 2016.
50. Brian Fisher, Tera Marie Green, and Richard Arias-Hernández. "Visual analytics as a translational cognitive science". In: *Topics in Cognitive Science* 3.3 (2011), pp. 609–625.
51. C. R. Fisher et al. "Efficient parameter estimation of cognitive models for real-time performance monitoring and adaptive interfaces". In: *Proceedings of the* 14th *International Conference on Cognitive Modeling (ICCM 2016)*. Ed. by D. Reitter and F. E. Ritter. University Park, PA, 2016.
52. John H Flavell. "Metacognition and cognitive monitoring: A new area of cognitive–developmental inquiry." In: *American psychologist* 34.10 (1979), p. 906.
53. Robert B Frary. "Formula scoring of multiple-choice tests (correction for guessing)". In: *Educational measurement: Issues and practice* 7.2 (1988), pp. 33–38. DOI: https://doi.org/10.1111/j.1745-3992.1988.tb00434.x.
54. Batya Friedman and Helen Nissenbaum. "Bias in computer systems". In: *ACM Transactions on Information Systems (TOIS)* 14.3 (1996), pp. 330–347.
55. Adrian Furnham. "The big five versus the big four: the relationship between the Myers-Briggs Type Indicator (MBTI) and NEO-PI five factor model of personality". In: *Personality and individual differences* 21.2 (1996), pp. 303–307.
56. Adrian Furnham and Hua Chu Boo. "A literature review of the anchoring effect". In: *The Journal of Socio-economics* 40.1 (2011), pp. 35–42.

57. Adrian Furnham, Joanna Moutafi, and Tomas Chamorro-Premuzic. "Personality and intelligence: Gender, the Big Five, self-estimated and psychometric intelligence". In: *International Journal of Selection and Assessment* 13.1 (2005), pp. 11–24.
58. R Michael Furr. *Psychometrics: an introduction*. SAGE publications, 2021.
59. Scott M Galster and Erica M Johnson. *Sense-assess-augment: A taxonomy for human effectiveness*. Tech. rep. AFRL-RH-WP-TM-2013-0002. Wright-Patterson AFB: Air Force Research Laboratory, 2013.
60. Gerd Gigerenzer and Wolfgang Gaissmaier. "Heuristic decision making". In: *Annual Review of Psychology* 62 (2011), pp. 451–482.
61. Stefan Gladisch, Heidrun Schumann, and Christian Tominski. "Navigation recommendations for exploring hierarchical graphs". In: *International Symposium on Visual Computing*. Springer. 2013, pp. 36–47.
62. Alix Goguey et al. "Interaction Pace and User Preferences". In: *Proceedings of the 2021 CHI Conference on Human Factors in Computing Systems*. Virtual Conference: ACM, pp. 1–14. DOI: https://doi.org/10.1145/3411764.3445772.
63. Lewis R. Goldberg. "A broad-bandwidth, public domain, personality inventory measuring the lower-level facets of several five-factor models". In: *Personality psychology in Europe* 7.1 (1999), pp. 7–28.
64. *Google Charts Introduction*. https://developers.google.com/chart/. Accessed: 2024-09-10.
65. Anthony G Greenwald, Debbie E McGhee, and Jordan LK Schwartz. "Measuring individual differences in implicit cognition: the implicit association test." In: *Journal of personality and social psychology* 74.6 (1998), p. 1464.
66. Peter E Hart, Nils J Nilsson, and Bertram Raphael. "A formal basis for the heuristic determination of minimum cost paths". In: *IEEE transactions on Systems Science and Cybernetics* 4.2 (1968), pp. 100–107.
67. Marti A Hearst. "Show It or Tell It? Text, Visualization, and Their Combination". In: *Communications of the ACM* 66.10 (2023), pp. 68–75.
68. Harold V Henderson and Paul F Velleman. "Building multiple regression models interactively". In: *Biometrics* (1981), pp. 391–411.
69. Andreas Hergovich, Reinhard Schott, and Christoph Burger. "Biased evaluation of abstracts depending on topic and conclusion: Further evidence of a confirmation bias within scientific psychology". In: *Current Psychology* 29 (2010), pp. 188–209.
70. Richards J. Heuer Jr. *Psychology of Intelligence Analysis*. Washington, DC, 1999.
71. Jessica Hullman, Eytan Adar, and Priti Shah. "Benefitting infovis with visual difficulties". In: *IEEE Transactions on Visualization and Computer Graphics* 17.12 (2011), pp. 2213–2222.
72. TJ Jankun-Kelly and Kwan-Liu Ma. "A spreadsheet interface for visualization exploration". In: *Proceedings of the Conference on Visualization'00*. IEEE Computer Society Press. 2000, pp. 69–76.
73. Daniel Kahneman. *Thinking, fast and slow*. Macmillan, 2011.
74. Daniel Kahneman and Shane Frederick. "A model of heuristic judgment". In: *The Cambridge Handbook of Thinking and Reasoning* (2005), pp. 267–294.
75. Alan S Kaufman. *Kaufman brief intelligence test: KBIT*. AGS, American Guidance Service Circle Pines, MN, 1990.
76. Daniel Keim et al. *Visual analytics: Definition, process, and challenges*. Springer, 2008. DOI: https://doi.org/10.1007/978-3-540-70956-5_7.
77. Dae Hyun Kim, Vidya Setlur, and Maneesh Agrawala. "Towards Understanding How Readers Integrate Charts and Captions: A Case Study with Line Charts". In: *Proceedings of the 2021 CHI Conference on Human Factors in Computing Systems*. 2021, pp. 1–11.
78. Hannah Kim et al. "Interaxis: Steering scatterplot axes via observation-level interaction". In: *IEEE transactions on visualization and computer graphics* 22.1 (2015), pp. 131–140.

79. Joshua Klayman. "Varieties of confirmation bias". In: *Psychology of learning and motivation* 32 (1995), pp. 385–418.
80. Ha-Kyung Kong, Zhicheng Liu, and Karrie Karahalios. "Frames and slants in titles of visualizations on controversial topics". In: *Proceedings of the 2018 CHI Conference on Human Factors in Computing Systems*. 2018, pp. 1–12.
81. Ha-Kyung Kong, Zhicheng Liu, and Karrie Karahalios. "Trust and recall of information across varying degrees of title-visualization misalignment". In: *Proceedings of the 2019 CHI Conference on Human Factors in Computing Systems*. 2019, pp. 1–13.
82. Ratanond Koonchanok et al. "Visual belief elicitation reduces the incidence of false discovery". In: *Proceedings of the 2023 CHI Conference on Human Factors in Computing Systems*. 2023, pp. 1–17.
83. Justin Kruger and David Dunning. "Unskilled and unaware of it: how difficulties in recognizing one's own incompetence lead to inflated self-assessments." In: *Journal of personality and social psychology* 77.6 (1999), p. 1121.
84. Bum Chul Kwon et al. "Axisketcher: Interactive nonlinear axis mapping of visualizations through user drawings". In: *IEEE transactions on visualization and computer graphics* 23.1 (2017), pp. 221–230.
85. Chufan Lai et al. "Automatic annotation synchronizing with textual description for visualization". In: *Proceedings of the 2020 CHI Conference on Human Factors in Computing Systems*. 2020, pp. 1–13.
86. Kibeom Lee and Michael C Ashton. "Psychometric properties of the HEXACO personality inventory". In: *Multivariate behavioral research* 39.2 (2004), pp. 329–358.
87. Peter Lee. *Learning from Tay's introduction*. https://blogs.microsoft.com/blog/2016/03/25/learning-tays-introduction/. Blog. 2016.
88. Sukwon Lee, Sung-Hee Kim, and Bum Chul Kwon. "Vlat: Development of a visualization literacy assessment test". In: *IEEE transactions on visualization and computer graphics* 23.1 (2017), pp. 551–560. DOI: https://doi.org/10.1109/TVCG.2016.2598920.
89. Paul Edward Lehner et al. "Confirmation bias in complex analyses". In: *IEEE Transactions on Systems, Man, and Cybernetics-Part A: Systems and Humans* 38.3 (2008), pp. 584–592.
90. Shiyao Li et al. "Confirmation Bias: The Double-Edged Sword of Data Facts in Visual Data Communication". In: *Proceedings of the 2025 CHI Conference on Human Factors in Computing Systems*. ACM, 2025.
91. Henry B Mann and Donald R Whitney. "On a test of whether one of two random variables is stochastically larger than the other". In: *The annals of mathematical statistics* (1947), pp. 50–60. DOI: https://doi.org/10.1214/aoms/1177730491.
92. Douglas Markant et al. "When do data visualizations persuade? The impact of prior attitudes on learning about correlations from scatterplot visualizations". In: *Proceedings of the 2023 CHI Conference on Human Factors in Computing Systems*. 2023, pp. 1–16.
93. Isaac D Mehlhaff. "A Group-Based Approach to Measuring Polarization". In: *American Political Science Review* (2022), pp. 1–9.
94. Mark Michael and Wen-Chun Lin. "Experimental study of information measure and inter-intra class distance ratios on feature selection and orderings". In: *IEEE Transactions on Systems, Man, and Cybernetics* 2 (1973), pp. 172–181. DOI: https://doi.org/10.1109/TSMC.1973.5408500.
95. Don A Moore and Paul J Healy. "The trouble with overconfidence". In: *Psychological review* 115.2 (2008), pp. 502–517. DOI: https://doi.org/10.1037/0033-295X.115.2.502.
96. Isabel Briggs Myers. "The Myers-Briggs Type Indicator: Manual (1962)." In: (1962).
97. Allen Newell. *Unified theories of cognition*. Harvard University Press, 1994.
98. Raymond S Nickerson. "Confirmation bias: A ubiquitous phenomenon in many guises". In: *Review of general psychology* 2.2 (1998), pp. 175–220.

99. Chris North et al. “Analytic Provenance: Process + Interaction + Insight”. In: *29th Annual CHI Conference on Human Factors in Computing Systems, CHI 2011* (2011), pp. 33–36. DOI: https://doi.org/10.1145/1979742.1979570. URL: http://www.scopus.com/inward/record.url?eid=2-s2.0-79957960481%7B%5C&%7DpartnerID=40%7B%5C&%7Dmd5=6e451badea5da358cb4d67c3595dfdf3.
100. Robert M Nosofsky. “The generalized context model: An exemplar model of classification”. In: *Formal Approaches in Categorization*. Ed. by Emmanuel M. Pothos and Andy J. Wills. Cambridge, UK: Cambridge University Press, 2011, pp. 18–39.
101. Robert M. Nosofsky. “Overall similarity and the identification of separable-dimension stimuli: A choice model analysis”. In: *Perception & Psychophysics* 38.5 (1985), pp. 415–432.
102. Stefan Palan and Christian Schitter. “Prolific. ac—A subject pool for online experiments”. In: *Journal of behavioral and experimental finance* 17 (2018), pp. 22–27.
103. Anshul Vikram Pandey et al. “The persuasive power of data visualization”. In: *IEEE transactions on visualization and computer graphics* 20.12 (2014), pp. 2211–2220.
104. Karl Pearson. “VII. Note on regression and inheritance in the case of two parents”. In: *proceedings of the royal society of London* 58.347-352 (1895), pp. 240–242. DOI: https://doi.org/10.1098/rspl.1895.0041.
105. Robert Pienta et al. “Scalable graph exploration and visualization: Sensemaking challenges and opportunities”. In: *2015 International Conference on Big Data and Smart Computing (BIGCOMP)*. IEEE. 2015, pp. 271–278.
106. William A. Pike et al. “The science of interaction”. In: *Information Visualization* 8.4 (2009), pp. 263–274.
107. Arthur E Poropat. “A meta-analysis of the five-factor model of personality and academic performance.” In: *Psychological bulletin* 135.2 (2009), p. 322.
108. Pieter Provoost. *Jerzy*. https://github.com/pieterprovoost/jerzy. 2017.
109. Eric D Ragan et al. “Characterizing provenance in visualization and data analysis: an organizational framework of provenance types and purposes”. In: *IEEE transactions on visualization and computer graphics* 22.1 (2016), pp. 31–40.
110. Donghao Ren et al. “Chartaccent: Annotation for data-driven storytelling”. In: *2017 IEEE Pacific Visualization Symposium (PacificVis)*. Ieee. 2017, pp. 230–239.
111. Brent W Roberts et al. “What is conscientiousness and how can it be assessed?” In: *Developmental psychology* 50.5 (2014), p. 1315. DOI: https://doi.org/10.1037/a0031109.
112. María Teresa Rodríguez, Sérgio Nunes, and Tiago Devezas. “Telling stories with data visualization”. In: *Proceedings of the 2015 workshop on narrative & hypertext*. 2015, pp. 7–11.
113. Neal J Roese. “Counterfactual thinking.” In: *Psychological bulletin* 121.1 (1997), p. 133.
114. Danielle J Sanchez and Ann Speed. *The Impact of Individual Traits on Domain Task Performance: Exploring the Dunning-Kruger Effect*. Tech. rep. Sandia National Lab.(SNL-NM), Albuquerque, NM (United States), 2020. DOI: https://doi.org/10.2172/1716582.
115. Ali Sarvghad, Melanie Tory, and Narges Mahyar. “Visualizing dimension coverage to support exploratory analysis”. In: *IEEE transactions on visualization and computer graphics* 23.1 (2016), pp. 21–30.
116. Edward Segel and Jeffrey Heer. “Narrative visualization: Telling stories with data”. In: *IEEE Transactions on Visualization and Computer Graphics* 16.6 (2010), pp. 1139–1148.
117. Samuel Sanford Shapiro and Martin B Wilk. “An analysis of variance test for normality (complete samples)”. In: *Biometrika* 52.3/4 (1965), pp. 591–611. DOI: https://doi.org/10.2307/2333709.
118. Ben Shneiderman. “The eyes have it: A task by data type taxonomy for information visualizations”. In: *The Craft of Information Visualization*. Elsevier, 2003, pp. 364–371.

119. Arjun Srinivasan et al. "Augmenting visualizations with interactive data facts to facilitate interpretation and communication". In: *IEEE transactions on visualization and computer graphics* 25.1 (2018), pp. 672–681.
120. Chase Stokes, Cindy Xiong Bearfield, and Marti A Hearst. "The Role of Text in Visualizations: How Annotations Shape Perceptions of Bias and Influence Predictions". In: *IEEE Transactions on Visualization and Computer Graphics* (2023).
121. Chase Stokes et al. "Striking a balance: Reader takeaways and preferences when integrating text and charts". In: *IEEE Transactions on Visualization and Computer Graphics* 29.1 (2022), pp. 1233–1243.
122. The Decision Lab. *Dunning–Kruger Effect.* https://thedecisionlab.com/biases/dunning-kruger-effect. Retrieved March 7, 2024. The Decision Lab, 2021.
123. Ayanna K. Thomas and Peter R. Millar. "Reducing the framing effect in older and younger adults by encouraging analytic processing". In: *The Journals of Gerontology Series B: Psychological Sciences and Social Sciences* 2.139 (2011).
124. Edmund R Thompson. "Development and validation of an international English big-five mini-markers". In: *Personality and individual differences* 45.6 (2008), pp. 542–548. DOI: https://doi.org/10.1016/j.paid.2008.06.013.
125. Edward R Tufte. "Envisioning information". In: *Optometry and Vision Science* 68.4 (1991), pp. 322–324.
126. Edward R Tufte and Peter R Graves-Morris. *The visual display of quantitative information.* Vol. 2. 9. Graphics press Cheshire, CT, 1983.
127. Amos Tversky and Daniel Kahneman. "Judgment under uncertainty: Heuristics and biases". In: *Science* 185 (1974), pp. 1124–1131.
128. Amos Tversky and Daniel Kahneman. "Rational choice and the framing of decisions". In: *Journal of Business* (1986), S251–S278.
129. Amos Tversky and Daniel Kahneman. "The framing of decisions and the psychology of choice". In: *Science* 211 (1981), pp. 453–458.
130. Kadri Umbleja, Manabu Ichino, and Hiroyuki Yaguchi. "Improving symbolic data visualization for pattern recognition and knowledge discovery". In: *Visual Informatics* 4.1 (2020), pp. 23–31. DOI: https://doi.org/10.1016/j.visinf.2019.12.003.
131. Susan VanderPlas and Heike Hofmann. "Spatial reasoning and data displays". In: *IEEE Transactions on Visualization and Computer Graphics* 22.1 (2016), pp. 459–468. DOI: https://doi.org/10.1109/TVCG.2015.2469125.
132. Dáša Vedejová and Vladimíra Čavojová. "Confirmation bias in information search, interpretation, and memory recall: Evidence from reasoning about four controversial topics". In: *Thinking & Reasoning* 28.1 (2022), pp. 1–28.
133. Emily Wall et al. "A formative study of interactive bias metrics in visual analytics using anchoring bias". In: *IFIP Conference on Human-Computer Interaction.* Springer. 2019, pp. 555–575.
134. Emily Wall et al. "A Markov Model of Users' Interactive Behavior in Scatterplots". In: *2019 IEEE Visualization Conference (VIS).* IEEE. 2019, pp. 81–85.
135. Emily Wall et al. "Left, Right, and Gender: Exploring Interaction Traces to Mitigate Human Biases". In: *arXiv preprint* arXiv:2108.03536 (2021).
136. Emily Wall et al. "Podium: Ranking Data Using Mixed-Initiative Visual Analytics". en. In: *IEEE Transactions on Visualization and Computer Graphics* 24.1 (Jan. 2018), pp. 288–297. ISSN: 1077-2626. DOI: https://doi.org/10.1109/TVCG.2017.2745078. URL: http://ieeexplore.ieee.org/document/8019863/ (visited on 04/09/2023).
137. Emily Wall et al. "Warning, bias may occur: A proposed approach to detecting cognitive bias in interactive visual analytics". In: *2017 IEEE Conference on Visual Analytics Science and Technology (VAST).* IEEE. 2017, pp. 104–115.

138. Arran Zeyu Wang, David Borland, and David Gotz. "A framework to improve causal inferences from visualizations using counterfactual operators". In: *Information Visualization* (2024), p. 14738716241265120.
139. Arran Zeyu Wang, David Borland, and David Gotz. "An empirical study of counterfactual visualization to support visual causal inference". In: *Information Visualization* 23.2 (2024), pp. 197–214.
140. Arran Zeyu Wang, David Borland, and David Gotz. "Beyond Correlation: Incorporating Counterfactual Guidance to Better Support Exploratory Visual Analysis". In: *IEEE Transactions on Visualization and Computer Graphics* 31.1 (2025).
141. Yun Wang et al. "Datashot: Automatic generation of fact sheets from tabular data". In: *IEEE transactions on visualization and computer graphics* 26.1 (2019), pp. 895–905.
142. Colin Ware. *Information Visualization: Perception for Design.* 2nd. San Francisco: Morgan Kaufmann Publishers Inc., 2004. DOI: https://doi.org/10.1016/B978-155860819-1/50001-7.
143. Eric W Weisstein. "Bonferroni correction". In: *https://mathworld.wolfram. com/* (2004).
144. Axel Westerwick, Benjamin K Johnson, and Silvia Knobloch-Westerwick. "Confirmation biases in selective exposure to political online information: Source bias vs. content bias". In: *Communication Monographs* 84.3 (2017), pp. 343–364.
145. Wesley Willett, Jeffrey Heer, and Maneesh Agrawala. "Scented widgets: Improving navigation cues with embedded visualizations". In: *IEEE Transactions on Visualization and Computer Graphics* 13.6 (2007), pp. 1129–1136.
146. Cindy Xiong, Lisanne van Weelden, and Steven Franconeri. "The curse of knowledge in visual data communication". In: *IEEE transactions on visualization and computer graphics* (2019).
147. Cindy Xiong et al. "Reasoning affordances with tables and bar charts". In: *IEEE transactions on visualization and computer graphics* (2022).
148. Cindy Xiong et al. "Seeing what you believe or believing what you see? belief biases correlation estimation". In: *IEEE Transactions on Visualization and Computer Graphics* 29.1 (2022), pp. 493–503.
149. Ji Soo Yi et al. "Toward a deeper understanding of the role of interaction in information visualization". In: *IEEE transactions on visualization and computer graphics* 13.6 (2007), pp. 1224–1231. DOI: https://doi.org/10.1109/TVCG.2007.70515.

Part III

Mitigate: Cognitive and Metacognitive Deficits

Asynchronous Interventions 6

6.1 Background: Intelligence Analysis

"Intelligence analysis" may conjure images of CIA agents and covert operatives, sifting through intercepted messages in time- and mission-critical settings of national security. Given the often very high stakes, it is of the highest priority to ensure that correct actions are taken. Clark [5] points out that there are two common culprits for failures in intelligence analysis: (i) flawed analysis or (ii) failure to act on analysis. This section focuses on the former—specifically, what strategies have been developed in the intelligence analysis community to prevent or mitigate flawed analysis?

A number of attempts have been made to mitigate bias in the domain of intelligence analysis, including training courses, videos, and reading material which can educate analysts about potential biases they may be susceptible to. While the goal is to promote informed decision-making by the analyst leading to a shift in user behavior, these techniques have not consistently proven to be effective. As articulated by Heuer:

> Cognitive biases are similar to optical illusions in that the error remains compelling even when one is fully aware of its nature. Awareness of the bias, by itself, does not produce a more accurate perception [10].

Alternative techniques that have proven marginally more effective include serious games [2, 6, 14, 22]. Serious games [21] are a class of games that have purposes beyond entertainment, including education, military, and healthcare applications. For instance, the MACBETH serious game [2] places the player in the perspective of an intelligence analyst tasked with identifying suspects, locations, and hypotheses about a potential terrorist attack. This video game trains players to recognize bias blind spot [18] (the tendency to value one's own knowledge and experiences over that of others'), fundamental attribution error [23] (the tendency to over-emphasize the role of personality characteristics while

E. Wall, *Human Bias in Visual Data Analysis*, Synthesis Lectures on Visualization,
https://doi.org/10.1007/978-3-032-09307-3_6

discounting situational factors), and confirmation bias [15] (the tendency to seek out evidence confirming a pre-existing hypothesis while dismissing or discounting disconfirming evidence). MACBETH did so through repeatedly quizzing players (bias blind spot); emphasizing situational cues in the game (fundamental attribution error); encouraging analysts to seek out disconfirming evidence, delay formalizing hypotheses, and explicitly generate alternative hypotheses through instructions (confirmation bias). Across three experiments, the researchers found mixed results on knowledge gains that decayed over time, and limited gains toward mitigation of the biases.

Alternatively, using a variation of the same MACBETH game, Dunbar et al. [6] found promising effects of serious games for mitigating confirmation bias [15] compared to alternative training videos. Collectively, research toward serious games suggests that they have potential to educate analysts about cognitive biases, but nonetheless do little to mitigate negative effects when biases inevitably occur in the analytic process. They reinforce that an analyst must be proactive using feedback to adjust their behaviors to mitigate the negative effects of bias, which we discuss more in Chap. 7.

6.2 Background: Workplace Bias

A number of specific biases have been identified as particularly concerning in the context of the workplace, hiring, and academic admissions. Implicit biases [8], for instance, shaped by cultural and societal norms, can perpetuate harmful societal stereotypes. The term "implicit bias" can include several more specific forms including examples such as racial and gender biases. This corresponds to "bias as a preconception" as described in Chap. 3.2. The implicit association test (IAT) described in Chap. 5.1 is used to measure these implicit biases.

Momentous research efforts have resulted in novel ways of quantifying these biases and produced numerous forms of empirical support to identify different forms of bias in the workplace. For instance, one laboratory experiment showed that science faculty exhibit a bias against female students such that male candidates are rated as significantly more competent and hireable than (otherwise identical) female candidates [13]. Similarly, racial discrimination has been studied in the labor market. Experimental results show significant discrimination against African-American names in job applications: equivalent resumes with White names receive 50% more callbacks for interviews than resumes with African-American names [1].

Specific to recommendation letters, which are often requisite components of applications in workplace and academic hiring, researchers have applied different methods to analyze gender bias using techniques such as qualitative content analysis [24, 25], word count-based analysis [7, 12, 20], and natural language processing techniques [19]. Zhang et al. applied qualitative analysis on letters of recommendation for applicants to a cardiology fellowship program and found that underrepresented applicants were more likely to be described using communal language and doubt raising language (hedging, faint praise, and negative language) [25]. To ascertain whether the language used in recommendation letters aligned with gender stereotypes of agency and communality, [12] used the Linguistic Inquiry and Word Count (LIWC) tool [4] to examine the word usage of the agentic word category (e.g., aggressiveness, assertiveness) and the communal word category (e.g., kindness, sympathy) in letters written for female and male applicants. The results suggested that women were more likely to be described as communal and less agentic than men. Similarly, Filippou et al. used LIWC to characterize the vocabulary in recommendation letters [7]. They discovered that male applicants were favored in terms of personal drive, work, and power, even among candidates with equivalent qualifications. Sarraf et al. utilized natural language processing techniques to extract the sentiment, emotion, and implicit tones conveyed by letter writers [19]. The findings revealed that recommendation letters for male applicants exhibited considerably more positive sentiment among female authors.

However, while there is an abundance of work *identifying* biases in workplace attitudes and application review practices, we observe significantly less work toward *mitigating* these biases, with more or less the same approaches applied in the workplace as in the intelligence analysis community. Attempts to mitigate implicit biases include training courses [9], often focusing on diversity initiatives [3] in organizational [16] and educational settings (e.g., to improve attitudes toward women in STEM [11] or toward individuals with disabilities in the workplace [17]). However, as with bias mitigation in the intelligence analysis community, these training efforts are likewise limited in impact to what a subject consciously remembers to apply during the decision-making process.

All of these efforts to mitigate biases, however, suffer from a similar limitation: that education of these biases requires that analysts *remember* information about biases and recall what strategies to apply to mitigate them. In reality, analysts may forget to engage the information in real-time decision-making scenarios. Thus in the next chapter, we review alternative techniques that operate in real time, as an individual engages in visual data analysis and decision-making.

References

1. Marianne Bertrand and Sendhil Mullainathan. "Are Emily and Greg more employable than Lakisha and Jamal? A field experiment on labor market discrimination". In: *American economic review* 94.4 (2004), pp. 991–1013.
2. Elena Bessarabova et al. "Mitigating bias blind spot via a serious video game". In: *Computers in Human Behavior* 62 (2016), pp. 452–466. URL: https://doi.org/10.1016/j.chb.2016.03.089.
3. Katerina Bezrukova, Karen A Jehn, and Chester S Spell. "Reviewing diversity training: Where we have been and where we should go". In: *Academy of Management Learning & Education* 11.2 (2012), pp. 207–227.
4. Ryan L Boyd et al. "The development and psychometric properties of LIWC-22". In: *Austin, TX: University of Texas at Austin* (2022), pp. 1–47.
5. Robert M Clark. *Intelligence analysis: a target-centric approach.* CQ press, 2019.

6. Norah E. Dunbar et al. "Mitigating cognitive bias through the use of serious games: Effects of feedback". In: *Lecture Notes in Computer Science (including subseries Lecture Notes in Artificial Intelligence and Lecture Notes in Bioinformatics)* 8462 LNCS (2014), pp. 92–105.
7. Pauline Filippou et al. "The presence of gender bias in letters of recommendations written for urology residency applicants". In: *Urology* 134 (2019), pp. 56–61.
8. Anthony G Greenwald and Linda Hamilton Krieger. "Implicit bias: Scientific foundations". In: *California Law Review* 94.4 (2006), pp. 945–967.
9. Diane F Halpern. "Teaching critical thinking for transfer across domains: Disposition, skills, structure training, and metacognitive monitoring." In: *American psychologist* 53.4 (1998), p. 449.
10. Richards J. Heuer Jr. *Psychology of Intelligence Analysis*. Washington, DC, 1999.
11. Sarah M Jackson, Amy L Hillard, and Tamera R Schneider. "Using implicit bias training to improve attitudes toward women in STEM". In: *Social Psychology of Education* 17.3 (2014), pp. 419–438.
12. Juan M Madera, Michelle R Hebl, and Randi C Martin. "Gender and letters of recommendation for academia: agentic and communal differences." In: *Journal of Applied Psychology* 94.6 (2009), p. 1591.
13. Corinne A Moss-Racusin et al. "Science faculty's subtle gender biases favor male students". In: *Proceedings of the national academy of sciences* 109.41 (2012), pp. 16474–16479.
14. Gwen Mullinix et al. "Heuristica: Decision a Serious Game for Improving Decision Making". In: *2013 IEEE International Games Innovation Conference (IGIC)* (2013), pp. 250–255.
15. Raymond S Nickerson. "Confirmation bias: A ubiquitous phenomenon in many guises". In: *Review of general psychology* 2.2 (1998), pp. 175–220.
16. Harold Andrew Patrick and Vincent Raj Kumar. "Managing workplace diversity: Issues and challenges". In: *Sage Open* 2.2 (2012), p. 2158244012444615.
17. Brian N Phillips et al. "Disability diversity training in the workplace: Systematic review and future directions". In: *Journal of occupational rehabilitation* 26 (2016), pp. 264–275.
18. Emily Pronin, Daniel Y Lin, and Lee Ross. "The bias blind spot: Perceptions of bias in self versus others". In: *Personality and Social Psychology Bulletin* 28.3 (2002), pp. 369–381.
19. Daniel Sarraf et al. "Use of artificial intelligence for gender bias analysis in letters of recommendation for general surgery residency candidates". In: *The American Journal of Surgery* 222.6 (2021), pp. 1051–1059.
20. Toni Schmader, Jessica Whitehead, and Vicki H Wysocki. "A linguistic comparison of letters of recommendation for male and female chemistry and biochemistry job applicants". In: *Sex roles* 57.7 (2007), pp. 509–514.
21. Tarja Susi, Mikael Johannesson, and Per Backlund. "Serious games: An overview". In: (2007).
22. Carl Symborski et al. "Missing: A Serious Game for the Mitigation of Cognitive Biases". In: *Interservice/Industry Training, Simulation, and Education Conference (I/ITSEC) 2014* (2014), pp. 1–13.
23. Philip E Tetlock. "Accountability: A social check on the fundamental attribution error". In: *Social psychology quarterly* (1985), pp. 227–236.
24. Frances Trix and Carolyn Psenka. "Exploring the Color of Glass: Letters of Recommendation for Female and Male Medical Faculty". en. In: *Discourse & Society* 14.2 (Mar. 2003), pp. 191–220. ISSN: 0957-9265, 1460-3624. DOI: https://doi.org/10.1177/0957926503014002277. URL: http://journals.sagepub.com/doi/10.1177/0957926503014002277 (visited on 04/29/2023).
25. Neil Zhang et al. "Race and gender bias in internal medicine program director letters of recommendation". In: *Journal of Graduate Medical Education* 13.3 (2021), pp. 335–344.

Real-Time Interventions 7

7.1 Background: Cognitive Interventions

One approach for mitigating cognitive biases in real-time is the use of non-technological strategies, including structured externalized processes, or structured analytic techniques [62]. Perhaps the most known and accepted is Analysis of Competing Hypotheses (ACH) [63]. ACH is a conscious tactic that can be used during the analytic process to evaluate the likelihood of multiple hypotheses in an unbiased way. It traditionally is implemented as a table, where each hypothesis is a column and each piece of evidence is a row. Each cell then indicates the extent to which the piece of evidence supports the given hypothesis. Hence, ACH creates a framework for analysts to assess the relevance of each piece of evidence for multiple hypotheses, and systematically eliminate less compelling hypotheses until a single most likely hypothesis remains. While an effective analytic tool, ACH is a time-consuming process not always used in practice.

Similarly, "consider the opposite" decision-making strategies make a thoughtful reflection of evidence for the alternative hypotheses a structured part of the decision-making process. This has shown promise to reduce some biases, including overconfidence, hindsight, and anchoring [10, 83]. However, these procedural thinking strategies come at the cost of potential cognitive overload, which can ultimately amplify some biases [97].

Using these techniques as inspiration, we can then consider how systems might *help* analysts in real-time to make better decisions. This opens up many questions surrounding how to most effectively mitigate the negative effects of cognitive bias:

- *How should the system inform the user when bias is detected?*
- *When and at what frequency should the system notify the user of bias or take initiative to intervene?*
- *How can a system minimize the cognitive burden on the user when bias is detected?*
- *To what extent should the system act on behalf of the user when bias is detected?*

E. Wall, *Human Bias in Visual Data Analysis*, Synthesis Lectures on Visualization,
https://doi.org/10.1007/978-3-032-09307-3_7

Thus, there is a rich space to be explored to understand the ways of intervening in biased decision-making processes.

7.2 Background: Machine Learning Interventions

Different forms of automated or technological support have been developed to address biases in computational systems at various stages of the machine learning pipeline: in the pre-processing stage [26, 44, 70, 127], in-processing stage [25, 27, 52, 89], and post-processing stage [5, 77].

In the pre-processing stage, the data re-weighing algorithm proposed by Kamiran et al. [70] generates weights for the training examples in each (group, label) combination differently to ensure fairness before classification. Similar to altering data weights, Friedler et al. [44] proposed Disparate Impact Remover to edit feature values to optimize the disparate impact directly between the unprivileged group (the group that suffers from model discrimination) and the privileged group.

At the in-processing stage of model training, novel modeling techniques can be employed to learn fair classifiers through approaches that aim to maximize accuracy while minimizing the ability to learn a protected variable (e.g., gender) so that predictions are made independent of the protected variable [128] or by adding discrimination aware regularization terms to an objective function that force a model to deal with differences in subgroups during training optimization [72]. Additionally, ensemble methods or focused learning can fit multiple classifiers to subgroups of data, thus ameliorating aggregation bias [16]. These algorithmic improvements aim to create a more fair model during training, even if the input data is biased.

Post-processing techniques change model predictions to improve fairness. Equalized odds postprocessing changes a model's inferences to make sure that for some protected attribute (e.g., sex, race) a model has equal performance for all of that attribute's values [59]; reject option classification helps increase the number of favorable outcomes assigned to underprivileged groups [71].

Importantly, although these techniques may improve the mathematical notion of model fairness, there are nonetheless numerous vulnerabilities for the introduction of human biases. Users may have to make decisions about whether to prioritize accuracy or fairness in some cases, which are subject to that person's bias if, for example, they are incentivized to create a model that is as accurate as possible. For feature engineering, some recent systems for human-in-the-loop ML have been designed to make recommendations that help users create new features [19] or interactively select features from different feature selection algorithms [73]. However, by making this process less automated, there is a tradeoff with the risk of introducing human biases that may affect feature quality.

Some efforts within the visualization community specifically have attempted to increase awareness on the human-front about the possible biases in their models. Cabrera et al.

proposed the FairVis system [24], where they introduce a visual analytic approach to monitor subgroup fairness in machine learning model results. Users can customize the dataset into multiple subgroups of interest, and then inspect the subgroup fairness under their own choices of metrics to highlight the under-performing subgroups. Microsoft proposed AI Fairness 360 [14], an interactive bias auditing system where users can choose their own fairness metrics and bias mitigation algorithms to apply to a dataset before model training, then visualize the fairness outcomes before and after applying the bias mitigation algorithm on the dataset. FairSight [6] proposed FairDM, a visualization system that communicates conventional fairness metrics, which can be integrated into the decision-making process. Wang et al. [123] proposed DiscriLens, an interactive visualization tool that facilitates an interpretation, evaluation, and comparison of algorithmic discrimination.

Collectively, these algorithms and systems can be used to intervene to mitigate biases in ML models.

7.3 Research: A Design Space of Real-Time Human Interventions

As interactive visualizations are increasingly used for data analysis and decision making in widespread domains, these processes can be improved by designing systems that can both leverage analysts' cognitive strengths and guard against cognitive limitations and weaknesses, including biases. In this section, we focus on **deriving a design space for visualization systems that can mitigate bias**. This work was originally published at IEEE VIS in 2019 [115].

Prior work detailing bias mitigation, or debiasing techniques (Chap. 6), has largely relied on non-technological strategies, like training courses [58, 63]. However, as data analysis increasingly takes place through technological media, particularly using visualization, we are motivated to consider ways in which visualization design can improve decision-making processes. While some prior work has provided guidelines toward mitigating one type of bias in a particular context [35, 51], we take a more general approach aimed at increasing real-time awareness of bias abstracted from a specific scenario. Given the recent emergence of bias mitigation in visualization (and hence relatively little work done in this area), our design space is derived from (1) prior work describing bias mitigation strategies outside of the visualization community, as well as (2) potential areas of visualization research that may inform the design of systems that mitigate bias.

Toward this goal, we must make a key assumption: that systems have information about bias in the user's decision or analytic process. Prior work has developed techniques that make this assumption reasonable. For example, computational methods exist for quantifying bias in the analytic process [51, 116, 117, 121] (see Chap. 5). Given this information, or other forms of de-biasing information, the goal is then to design systems that can help people make better decisions by compensating for the ways in which people are likely to make cognitive errors.

Many different types of biases can have common observed behavioral effects [74, 121]. For example, an analyst subject to vividness criterion [63] (over-reliance on information that is vivid or personal) may interact with a particularly vivid data point repeatedly. The same behavior is likely to be observed if the analyst is subject to a different type of cognitive bias, like the continued influence effect [63] (continued reliance on a piece of evidence, even after it has been discredited). As a result, some mitigation strategies can, to varying degrees, have an effect on multiple types of biases [10]. Hence, in this design space, we do not focus on any specific type of cognitive bias. Rather, we find it prudent to introduce design considerations for mitigating bias and improving analytic *processes*, agnostic to a specific type of bias.

Within this context, the contribution of this work is the derivation of 8 dimensions of visualization design that designers should consider when developing systems to mitigate biased decision making, or retrofitting such capabilities in existing tools. These dimensions represent aspects of a visualization system that can be manipulated in specific contexts to mitigate biased decision making. We concretize these dimensions through examples using a hypothetical visual analytic system, `fetch.data`, to illustrate potential bias mitigation interventions. While prior work on bias mitigation in the context of visualization and visual analytics is limited, we find it timely to scaffold design efforts going forward when building systems that can mitigate biased decision making.

7.3.1 Driving Areas in Visualization Research

While prior work on mitigating cognitive bias in the visualization domain is sparse (e.g., [35, 51, 119]), we are motivated to define a design space in this emergent area. Hence, in the context of visualization research, we derive inspiration from sub-fields of visualization research that may be leveraged to mitigate biased decision-making processes.

Guidance

According to Ceneda et al., guidance can be defined as "a computer-assisted process that aims to actively resolve a knowledge gap encountered by users during an interactive visual analytics session" [28]. In other words, systems that guide users provide some form of assistance during interactive data analysis (e.g., Scented Widgets [125] in collaborative settings or VizAssist [17] for visualization creation).

Bias mitigation in visual analytics can be loosely thought of as a form of guidance, where the goal is to impact the user's decision making in such a way as to promote a more balanced analytic *process* and/or a more reasonable *product* or final choice decision. Within Ceneda et al.'s [28] characterization of guidance, we focus on the *output means* in the context of bias mitigation. What can we show the user to facilitate an analytic process that is less prone to the potentially negative effects of cognitive bias?

Analytic Provenance and Awareness

Analytic provenance is a description of the analytic process leading to a decision [87]. Many researchers have shown the impact of raising users' awareness of their process. Researchers have shown ways to measure or visualize the user's *coverage* of the data throughout analysis [11, 68], leading users to make more discoveries [125] and analyze the data more broadly [42, 75]. This body of research shows promise that provenance awareness can alter user behavior in the context of bias mitigation. In upcoming Sects. 7.4–7.5, we describe recent research efforts that explicitly employ this concept to develop systems (Sect. 7.4) and empirical findings (Sect. 7.5) in support of these techniques for bias mitigation.

We posit that there are many other ways visualization and visual analytic tools can mitigate bias in real-time by simply increasing the user's awareness of their process. For example, Dimara et al. [35] observed that highlighting points in a scatterplot that fall on the Pareto front led to a lower susceptibility to the attraction effect. However, we can envision mitigation strategies in a variety of interactive settings. One approach could be to characterize the analysis process and alert the user of detected biases in real-time in the form of notifications. There is a rich space of prior work around notifications, including the tradeoffs of push versus pull [22, 79], the effects of timing and interruptions [4, 32], etc. that can inform the design of notification systems for increasing user awareness of biased analytic processes. In fact, more recently, Narechania et al. developed a library that overlays provenance on visualizations with a set of reusable widgets [85].

Mixed-Initiative Visual Analytics

Mixed-initiative [65] visual analytic tools explore the balance between human and machine effort and responsibilities. Some systems leverage users' interaction sequences to infer about their goals and intentions in an analytic model (e.g., [21, 120]). These types of mixed-initiative tools inspire potential ways of mitigating cognitive bias as people use visualizations. Some mixed-initiative efforts have already begun to integrate visual analytic recommendations based on user interest or semantic interactions [38]. Gladisch and colleagues [50] even suggest using the notion of interest through user interactions to penalize users or down-weight some recommendations to guide the user to other parts of the data space. By considering the spectrum of machine initiative, we can ideate novel solutions to steer users around bias-related pitfalls. In particular, the machine could operate as an unbiased collaborator that can act on behalf of the user, or *take initiative*, to mitigate biased analysis processes.

Collaboration

Feedback about biased behaviors can be given to a third party agent (e.g., a human teammate or a supervisor) to leverage "wisdom of crowds" [78, 105] to cancel out the noise of potentially sub-optimal individual decisions [64, 90]. This strategy could prove useful in collaborative analytic settings. For example, analysts teaming on a project may be alerted to each other's biased behaviors, to ensure they cross-validate each other's work. In this case, prior work on fostering awareness in collaborative settings can be informative [11–13, 56, 57, 107]. The rutaBAGA system described in Sect. 7.6 leverages collaboration to mitigate biases in university admissions.

Interface and Interaction Design

Law and Basole [75] describe interface design considerations for encouraging more broad data exploration as a means for mitigating some biases. Design considerations include what constitutes a unit of exploration (i.e., to track breadth of exploration across data points or dimensions), the distinction between user- and system-driven exploration (i.e., to focus on showing users meaningful information about their own analytic process or the system's computational assumptions), and the alternatives of related and systemic exploration (i.e., whether the expansion of information is driven by the user or not). They demonstrate a prototype tool for social network analysis that considers these dimensions in the design of the interface for encouraging broad exploration of the data. Similarly, Sukumar and Metoyer [104, 106] present design considerations for bias mitigation in the context of the college admissions process. Their guidelines include a broad focus on tasks like easing cognitive load, supporting sensemaking, de-correlating error (i.e., not viewing students' applications as a whole), mobilizing system 2 (i.e., encouraging thoughtful consideration of application content over quick intuitive judgments), and combining formulas with intuition (i.e., to limit the drawbacks of either in isolation).

A recent study by Dimara et al. [35] showed that interaction design can itself serve to mitigate bias (namely, the attraction effect). The attraction effect describes the phenomenon where a person's decision between two alternatives is altered by the introduction of an irrelevant third option. In a decision-making task using scatterplots, users were required to utilize a process of elimination, clicking to remove individual points on the scatterplot, until a single point remained as their final decision. With this interaction design, researchers observed a reduction in the attraction effect compared to participants who simply clicked on their final decision point (without first eliminating others). Hence, considering alternative interaction designs for a task could serve as an effective intervention for mitigating biased decision making.

7.3.2 Design Space

In this section, we describe 8 dimensions (D1-D8) important to the design of bias mitigation strategies in VA tools. These 8 dimensions are not strictly orthogonal, nor are they exhaustive. Rather, they represent our view on the aspects of visualization systems that may be manipulated for the purposes of mitigating bias given current technologies. Due to limited prior work on bias mitigation in VA, the process for deriving this design space was largely ad-hoc, guided primarily by literature review in driving areas of vis research (Sect. 7.3.1). Many of the dimensions are related (Fig. 7.1).

To ground our design space, we describe applied examples using a common scenario. Suppose a hiring manager at a tech company uses a visual analytic tool, `fetch.data` (Fig. 7.2, top) to analyze tabular data about job applicants. From potentially hundreds of applications on file, the hiring manager wants to select a handful of candidates to interview. Suppose the system is comprised of three interactive views: (A) a scatterplot view, (B) a filter panel, and (C) a ranking table view. Scatterplot axes can be configured, the table sorted, and filters used to adjust the subset of data viewed. For each design dimension below, we describe the concept and revisit this example to illustrate how a visualization could be retrofitted to mitigate biased decision making.

D1: Visual Representation

Concept. There are many possibilities for representing information that may have a debiasing effect. For bias interventions intended not to impose significant disruption to the user's natural analytic process, designers may opt for *peripheral* or ex-situ visualizations. Peripheral visualizations would appear in a separate view of the interface, potentially available on demand, and hence may be less likely to call the user's attention away from the primary visualization. On the other hand, in-situ visualizations would appear within existing views. For example, in-situ visualizations could encode debiasing information in previously unused visual channels (e.g., opacity, color, position, etc.). The choice between in-situ versus peripheral display of debiasing information should be informed by (1) the type of debiasing information, and (2) the intended level of user attention to that information. Furthermore, the representation of information should follow conventions described in vis research. For example, chart types [96] and optimal visual encodings [30] should be considered based on the type of data presented.

Example. Suppose the hiring manager is subject to anchoring bias, or the tendency to rely too heavily on initial "anchoring" information [40] (in this case, the first few résumés received). If the first handful of candidates happened to be males, successful bias mitigation strategies could draw the user's attention away from potential gender bias. Some metrics of bias (e.g., those described in Sect. 5.2) compare the distribution of user interactions to the underlying distributions of the data. This could be shown in a *peripheral* view showing both distributions (Fig. 7.2, B.2). Alternatively, a single metric quantifying the severity of the bias could be encoded as an ambient background display where color or opacity represents level of bias. In another example, history (provenance) could be shown in-situ by encoding size of scatterplot points as time spent examining each candidate, drawing attention to those (female candidates) who may have been unintentionally ignored (Fig. 7.2, A.1). The Lumos visualization system, described in Sect. 7.4, uses both in-situ and peripheral displays of bias [84].

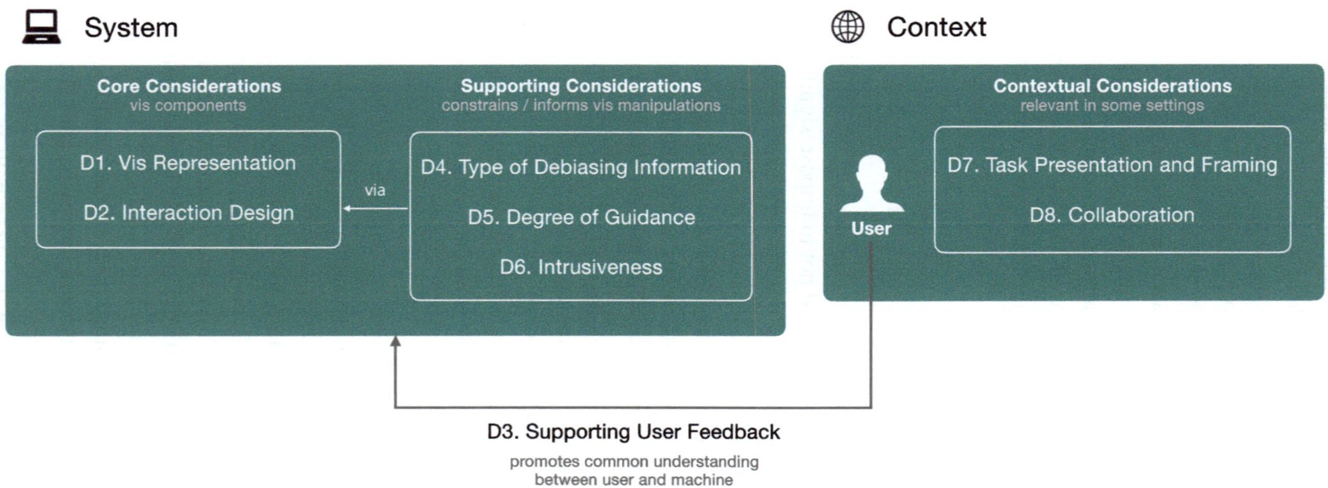

Fig. 7.1 The design space is comprised of 8 dimensions, described in Sect. 7.3.2. D1 (VISUAL REPRESENTATION) and D2 (INTERACTION DESIGN) are the two *core* components of a visualization [43] that can be manipulated to mitigate biased decision-making processes. How these components are manipulated is informed and constrained by *supporting* considerations, including D4 (TYPE OF DEBIASING INFORMATION), D5 (DEGREE OF GUIDANCE) and D6 (INTRUSIVENESS). Some *contextual* considerations may only be relevant in specific settings, including D7 (TASK PRESENTATION AND FRAMING) and D8 (COLLABORATION). Finally, D3 (SUPPORTING USER FEEDBACK) connects the user and contextual setting to the system by promoting a common understanding between user and machine

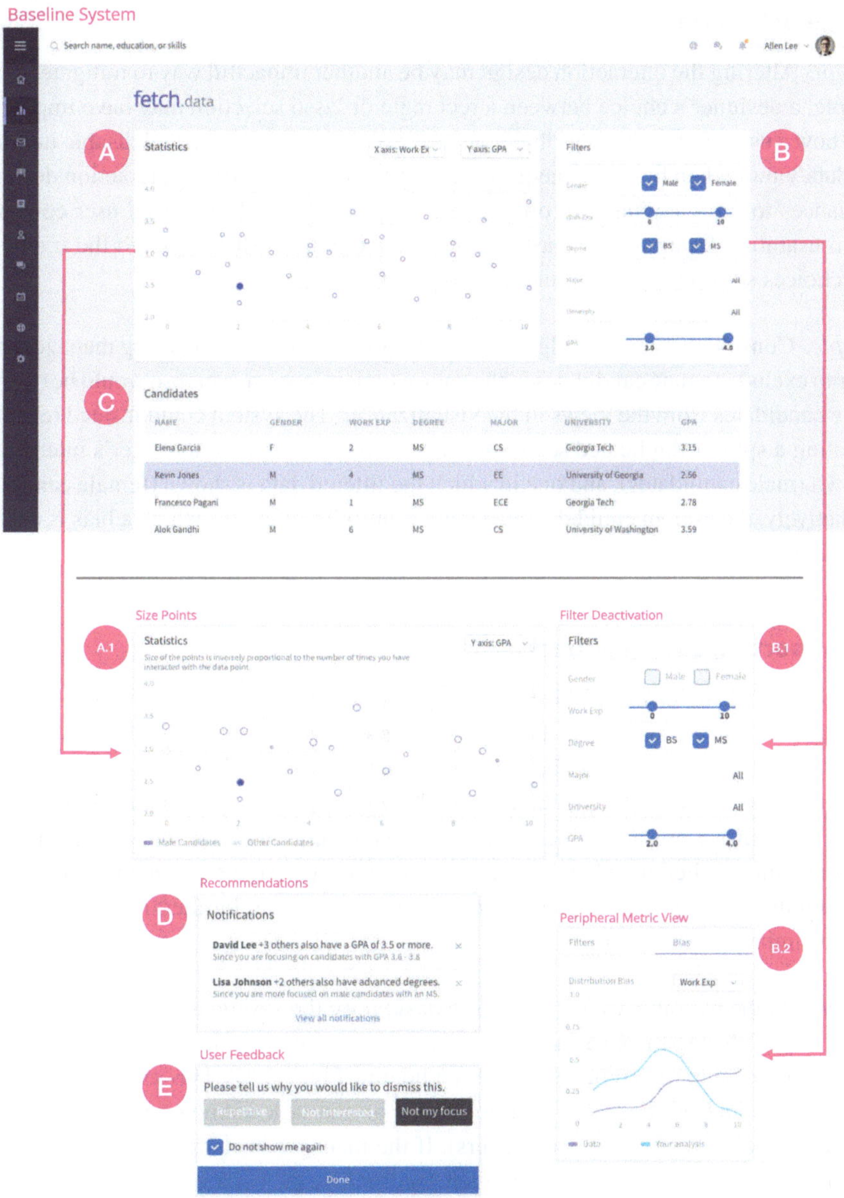

Fig. 7.2 An illustration of the system, `fetch.data`, used to analyze tabular data about job applicants. The baseline system (top) consists of (A) a scatterplot view, (B) a filter panel, and (C) a ranking table view. Possible bias mitigation interventions are shown below. (A.1) sizes candidate data points based on the analyst's interactions with them. (B.1) shows the system disabling the gender filter. (B.2) shows a peripheral view of metrics quantifying bias. (D) shows recommendations for candidates the analyst has not yet examined. (E) shows a pop-up allowing the analyst to provide feedback when dismissing a notification

D2: Interaction Design

Concept. Altering the interaction design may be another impactful way to mitigate bias. For example, a designer's choice between a rectangle or lasso selection may have implications about how a user approaches a problem or task. Similarly, a system could disable interaction with data/views when biased behavior is detected. However, altering interaction design and affordances to mitigate bias can often come at the expense of perceived user control and system usability. Designers of bias mitigation interventions should weigh the tradeoffs of these choices so usability is not unduly compromised.

Example. Consider a filtering widget designed to mitigate bias. If the hiring manager applies a filter to exclude female candidates in the data, a typical system response would be to remove female candidates from the views in the visualization. The system could instead respond by presenting a split or duplicated scatterplot view: one in which the manager's intended data is shown (male candidates), and one in which the filtered data is shown (female candidates). Alternatively, the system could disable interactions with filters for which a bias is exhibited (Fig. 7.2, B.1).

D3: Supporting User Feedback

Concept. While the primary objective of bias mitigation interventions is to communicate information from the system to the user, supporting user feedback is likewise important. In real-world systems that may be able to characterize user bias with limited accuracy, it can enable the user to communicate information outside the scope to the underlying model of bias (e.g., that a presumed bias is not due to unconscious error, but rather an external task constraint). When user feedback is supported, users may be given an increased sense of mutual understanding or common ground with the system. Further, models of user bias might be improved as a result.

Example. In the hypothetical hiring scenario, suppose the system detects a strong (gender) bias in that the hiring manager has primarily interacted with male candidates. One system response could be to recommend female candidates (Fig. 7.2, D). However, the hiring manager's focus could be the result of a constraint on the task unknown to the system (e.g., a division of labor between two managers). If the manager dismisses the recommendation of female candidates, the system can elicit feedback (e.g., via a pop-up dialog) to clarify information potentially outside the system's purview (Fig. 7.2, E). Reasons may include things like a repetitive recommendation, an irrelevant recommendation, or an external task constraint. According to the hiring manager's selection, the system may alter the underlying model of bias to account for these preferences or constraints.

D4: Type of Debiasing Information

Concept. A primary consideration in designing bias mitigation strategies is the type of debiasing information that the system will capture and communicate to the user. Types of debiasing information that could promote user awareness includes things like analytic provenance, summative metrics that quantify the analytic process [41, 67], and so on. We could further conceive of future systems that are able to identify specific types of bias the user may be subject to by name (e.g., confirmation bias [86], anchoring bias [40], etc.). Systems should ideally communicate information about potential biases in a way that guides users to counteract them (i.e., they should be *informative* and *actionable*).

Example. Suppose the hiring manager is exhibiting signs of availability bias [111], or a heavy reliance on information that is most easily remembered or most recent (i.e., the most recent application received). When bias is detected (i.e., the hiring manager is exhibiting signs of availability bias), the system could show provenance information to the hiring manager by adding an additional view to the interface that shows a snapshot of various stages of history of the manager's analytic process (e.g., like the history shown in [61]). Alternatively, the system could show the results of summative interaction metrics, similar to the metric visualization in [121] (Fig. 7.2, B.2). This could enable the hiring manager to reflect on their process and adjust.

D5: Degree of Guidance

Concept. Degree of guidance is analogous to Ceneda et al.'s *guidance degree* in the context of visual analytic system guidance [28]. It can be thought of as a spectrum that refers to how much the system "helps" the user. On one end of the spectrum, the system provides little intervention, while on the other end, the system more aggressively steers the user. Ceneda et al. describe three scenarios for degrees of guidance: *orienting*, *directing*, and *prescribing*, examples of which are described below. The degree of guidance adopted must be considered alongside tradeoffs of user experience. Systems that deny user control may come at the expense of perceived usability issues.

Example. An *orienting* bias mitigation strategy would promote user awareness of their biases. For example, the system could size candidates in the scatterplot according to the hiring manager's focus (where larger points represent neglected candidates; Fig. 7.2, A.1). A *directing* bias mitigation strategy could suggest candidates to the hiring manager to consider from the pool of candidates who have not been analyzed (Fig. 7.2, D). A *prescribing* bias mitigation strategy would involve the system assuming initiative or otherwise taking control from the user. An example of this might be disabling filters or interactions with specific candidates (Fig. 7.2, B.1).

D6: Intrusiveness

Concept. Intrusiveness refers to how much the system interrupts or otherwise intrudes on the user's analysis process. On the low end of the spectrum, bias information may be presented peripherally or even on demand (i.e., user attention optional). Highly intrusive mitigation strategies may present information front and center requiring the user's attention until the perceived bias is addressed. The level of intrusiveness of the intervention should not outweigh the intended benefit, however. In lower-cost decisions (e.g., analyzing a dataset of food to construct a weekly menu), a highly intrusive bias mitigation strategy would likely be unwelcome to the user. On the other hand, the intrusion may be acceptable for decisions that carry greater importance (e.g., criminal intelligence analysis). This dimension is distinct from D5 (Degree of Guidance). Consider the following analogy: suppose a person asks her friend for directions from point A to point B. The friend may draw a map, suggest GPS, or walk her friend there herself (i.e., degree of guidance). If she walks with her friend, she may exhibit a spectrum of intrusiveness (e.g., how closely does she stand to her friend).

Example. In our hypothetical scenario, a minimally intrusive mitigation strategy may present bias information to the hiring manager only *on-demand*. For example, there may be a tab in the interface that reveals information about the model of user bias when clicked on (Fig. 7.2, B.2). A more intrusive bias mitigation strategy could be a pop-up notification that repeatedly alerts the hiring manager until a less biased analysis state is reached (Fig. 7.2, D).

D7: Task Presentation and Framing

Concept. Changes to the presentation of information can have an impact on the analytic process and outcome. Framing has been found to strongly shape decision-making [109], including richness of language used and positive versus negative terminology to describe logically equivalent information [112]. For instance, in one study, researchers showed that people chose one treatment (surgery) over another (radiation therapy) when it was described as having a 90% short-term survival rate versus a 10% immediate mortality rate [80]. In addition to language, visual framing or anchoring can also shape decision making [29]. In situations where designers of bias mitigation interventions have control of the task, thoughtful consideration should be given to the often subtle-seeming aspects of task presentation.

Example. This contextual consideration is primarily limited to situations in which the designer has control over the presentation of the task. In our hypothetical scenario, the job description can impact the analysis process. For example, the framing of criminal background criteria may alter the hiring manager's threshold for minimally viable candidates (e.g., the negative framing "does not have a criminal record" may lead to a lower decision threshold than the positive framing "has a clean record"). Visual framing of information can also impact decision making (i.e., the relative size and spatial arrangement of multiple views, the order in which the hiring manager is trained to use them, etc.).

D8: Collaboration

Concept. Collaborative contexts have potential to mitigate bias by allowing others to check an analyst's work. By leveraging "wisdom of crowds", collaboration helps to ensure that a single sub-optimal individual decision does not prevail [64, 90]. Analysts teaming on a project may be alerted to biased behaviors, to ensure they cross-validate each other's work. In this case, prior work on fostering awareness in collaborative settings can be informative [11, 13, 57]. Collaboration is contextually relevant, as it may be infeasible in many scenarios due to the nature of the decision (e.g., a personal healthcare decision) or other constraints (e.g., division of labor).

Example. To leverage collaboration to mitigate biased decision making, designers of the vis tool could show traces of other hiring managers' exploration behaviors. For example, this could entail coloring points in a scatterplot based on which have been previously examined by other hiring managers (e.g., [11]), to promote social accountability.

7.3.3 Characterizing Existing Systems

Two recent works have designed interventions within visualization systems to mitigate cognitive bias [35, 51]. For each, we describe the context of the problem, the bias intervention, and how it fits within the aforementioned design space.

Mitigating Selection Bias

In analyzing high dimensional datasets, many dimensions may exhibit correlations. For instance, a dataset of patients to be selected for a study on a new medication may show correlations between dimensions like cataracts and age of patients. Hence, when attempting to select a sample from a larger dataset, the analyst may unintentionally filter out a representative part of a population (i.e., selection bias) [51].

To mitigate selection bias, Gotz et al. modified a visualization tool, DecisionFlow. Specifically, they modified an existing view in the visualization (D1, in-situ) by adding a color-coded bar after each subsequent data selection to depict the similarity of the subset to the original dataset. The color-coding of the bar was based on a computed value (D4, bias metric) that quantified the differences in variable distributions between the two datasets. They also added a secondary view (D1, ex-situ) that provided details about how variables of the data were constrained either via direct or unintentional filtering via correlation. These modifications represent an *orienting* degree of guidance (D5) that is relatively unintrusive (D6). They did not modify the interaction design (D2), task presentation (D7), or collaborative nature (D8) of the system, and did not enable user feedback (D3).

Mitigating the Attraction Effect

Dimara et al. designed an experiment to test two different strategies for mitigating the attraction effect (the phenomenon where a person's decision between two alternatives is altered by the introduction of an irrelevant third option) in scatterplots [35].

In one strategy, they highlighted optimal choices with a brightly colored stroke (D1, in-situ) before users clicked to select their choice point. This constitutes an *orienting* degree of guidance (D5). In another design, they altered the task framing (D7) and interaction design (D2) from "select a point" to "eliminate points until only one remains". While this was more effective than the first strategy, it could have usability implications as it represents a more intrusive (D6) design. For both strategies, they do not support user feedback (D3) or collaboration (D8). By virtue of these mitigation strategies taking place within an experiment, the debiasing information (D4) was a precondition to the study.

7.4 Research: Lumos—A System for Promoting Awareness of Analytic Behaviors

Unconscious biases can take many forms, some of which are relatively innocuous (i.e., personal preference for a particular genre of movie) while others can lead to costly incorrect decisions or engender harmful societal stereotypes. For instance, dark-skinned people are denied parole (racial bias) [9], women experience a variety of barriers to equity in C-suite promotions (gender bias) [69], ailing but younger people are denied optimal treatment (age bias) [99], etc. These are instances of "bias as a preconception" from Sect. 3.2. Even when these factors are not actively considered, they can implicitly affect the way people examine and process new information [55], and hence their analytic behaviors.

Apart from implicit biases and stereotypes, there are other cognitive and perceptual biases that also influence people's analytic behaviors. Cognitive biases, in particular, describe systematic errors that can result from the use of "fast and frugal" heuristics [49] to make decisions ("bias as a cognitive processing error" from Sect. 3.2). Several biases have been shown to extend to tasks involving visualizations for decision making (e.g., [29, 34, 114, 116]). Yet, common visual data analysis tools such as Tableau and Microsoft Excel that help users see and understand their data do not report analytic behaviors that may correspond to such biases. Therefore we ask: how much can understanding data analysis and decision-making behaviors reduce the potentially negative influences of cognitive, perceptual, or societal biases, if users were simply more aware of these often unconscious factors?

Originally published at IEEE VIS in 2021 [84], we present Lumos, an analysis tool that visualizes interaction history with data (i.e., **interaction traces** [119]) to increase **awareness** of potential interaction biases that influence data analysis and decision-making processes. Using in-situ and ex-situ visualization techniques, Lumos provides real-time feedback about a user's **analytic behavior** for self-awareness and self-reflection to potentially change future course. For example, Lumos remembers and highlights data points that have been previously examined in the same visualization (in-situ) and overlays the interacted data points on the underlying data distribution in a separate visualization (ex-situ) for comparison. Furthermore, Lumos allows users to configure a custom target distribution to reflect decision-making goals (e.g., a university admissions committee in a computer science department may define an analysis target of 60% female to promote increased gender diversity in the department, even if only 40% of applicants are female). We posit that Lumos can improve behaviors exhibited during data exploration and decision-making to help mitigate the dangers of human interaction biases affecting judgments and foster more transparent analysis processes.

The primary contributions of this work include **(1)** a technique to model users' **analytic behavior** from interactions with the data, **(2)** a visual analysis tool, Lumos (https://lumos-vis.github.io/), that implements visualizing analytic behavior using interaction traces, **(3)** a series of scenarios that describe potential usage of Lumos, and **(4)** results from a user study to understand how Lumos helps users be more aware of their analytic behavior during visual data analysis.

7.4.1 Lumos

In this section we first define the terminologies used in this section, enumerate our design goals, and describe the Lumos user interface.

Interaction Logs. Telemetry data of a user's interactions (e.g., hovers) with the user interface elements (e.g., data points on a scatterplot).

Analytic Behavior Model. A model that quantifies user's behavior (set of actions) from interaction logs (e.g., computing the distribution of interacted data points and attributes).

Interaction Traces. Visual feedback of the user's analytic behavior in the user interface; this is shown by adding visual scents in two ways: in-situ (at the place of interaction) and ex-situ (in an external view).

Awareness. Knowing what is going on [39] or what has been done and found during the exploration process to perform effective reasoning [101]; in this context, knowledge gained from inspecting one's analytic behavior via interaction traces.

Design Goals

Our development of Lumos was driven by **five** key design goals. We compiled these goals based on a combination of similar prior visual analysis tools [42, 51, 98], formative feedback from pilot studies, and our own hypotheses with respect to usability.

DG1. Capture and present analytic behavior with attributes. Overemphasis (or underemphasis) on specific attributes during data exploration may lead to unconscious biases (e.g., not interacting with a Gender attribute may practically result in a bias toward men if the dataset has more men than women). This goal translates to capturing user interactions with *attributes*, modeling analytic behavior, and showing interaction traces to increase awareness to influence changes in subsequent interactions.

DG2. Capture and present analytic behavior with data points. Overemphasis (or underemphasis) can also occur on specific values of data (e.g., interacting mostly with a few top candidates for university admissions may come at the expense of neglecting other candidates). This translates to the same goal as **DG1** but at the data point-level.

DG3. Facilitate configuring different target distributions. Determining overemphasis (or underemphasis) on specific attributes or data requires comparing a user's analytic behavior with a known target distribution (e.g., the underlying data) as a baseline. However, different domains, tasks, attributes, or social norms may call for different target distributions. This goal translates to allowing users to configure different target distributions to suit their requirements.

DG4. Facilitate comparison between analytic behavior and a baseline distribution. This goal translates to visualizing the user's analytic behavior and the configured target distribution and quantifying the difference (or similarity) between the two distributions.

DG5. Facilitate visual data exploration while showing awareness. This goal ensures that visual data exploration usability is not sacrificed by the added awareness visualization techniques.

Quantifying Analytic Behavior

We quantify *analytic behavior* using (1) the attribute distribution (AD) metric [121] and (2) relative frequency of interactions with data and attributes. The AD metric, along a scale from 0 (no bias) to 1 (high bias), characterizes how a user's interactive behavior deviates from expected behavior. By default, the system chooses a **proportional** baseline of expected behavior, wherein interactions with any given data point are equally likely, reflecting the true underlying distributions of attributes in the dataset. For example, if a user interacts primarily with PG-13 movies among a dataset of movies that contains predominantly G-rated movies, the AD metric for the Content Rating attribute will be high (more emphasis). If the user instead spent more time interacting with G-rated movies, proportional to the distributions in the dataset, the AD metric value for Content Rating would be low (less emphasis). Alternative baselines can be set using Lumos (**DG3**).

User Interface

The Lumos user interface consists of the following views:

(A) **Data Panel** shows the currently loaded dataset.

(B) **Attribute Panel** shows a list of attributes in the dataset along with their datatypes: {Nominal (N; **A**), Quantitative (Q; **#**), Temporal (T; 📅)} and buttons to apply a filter (▼).

(C) **Encoding Panel** shows UI controls (dropdowns) to create visualizations by specifiying different encodings: {Chart Type, X Axis, Y Axis, Aggregation}. Lumos currently supports *four* visualization types: *{scatterplot, strip plot, bar chart, line chart}* and *five* aggregation types: *{count, sum, minimum, maximum, average}* depending on the attribute data type combinations.

(D) **Filter Panel** shows UI controls (range sliders for {Q, T} and multiselect-dropdowns for {N} attributes) to filter data. Filters can be added by clicking on ▼ in the Attribute Panel (B) **(DG5)**.

(E) **Visualization Canvas** renders the visualization based on the Encoding (C) and Filter (D) Panel specifications.

(F) **Details View** shows additional information when the visualization elements (e.g., point, bar, strip) in (E) are interacted with. Hovering on a single data point (e.g., a strip in a strip plot) shows a list of all attribute values for the given data point (Fig. 7.3F). Hovering on an aggregation of data points (e.g., a bar of a bar chart) shows a table of all data points that belong to that aggregation (the bar) with attributes as columns and values as rows (Fig. 7.4).

(G) **Distribution Panel** shows a list of attribute cards similar to the Attribute Panel where clicking on a card toggles open/close a visualization that overlays user's interaction traces on data points (blue area) on the target distribution (black curve) **(DG3)**.

In addition to these features, Lumos supports additional customization options (⚙): (1) *Sort By* to sort attributes in the Attribute panel by their {Order-in-Dataset (default), Name, Datatype, Focus} (2) *Color Mode* to determine the normalization strategy to compute the interaction frequency: *Relative* (default) divides each value by the maximum of all values; *Absolute* divides each value by the sum of all values; *Binary* treats no interactions=`0` and at least one interaction=`1`, (3) *Focus Mode* that determines the Y-encoding in the Distribution Panel visualizations: *Percentage* (default) shows the percentage of Focus and Target distributions whereas *Raw* shows the absolute counts; and (4) *Color Scale* to choose different color scales (e.g., Sequential, Divergent).

Based on user interaction logs with attributes and data points, Lumos computes analytic behavior and presents real-time interaction traces back to the user. Analytic behavior and interaction traces are observed in multiple ways in the interface.

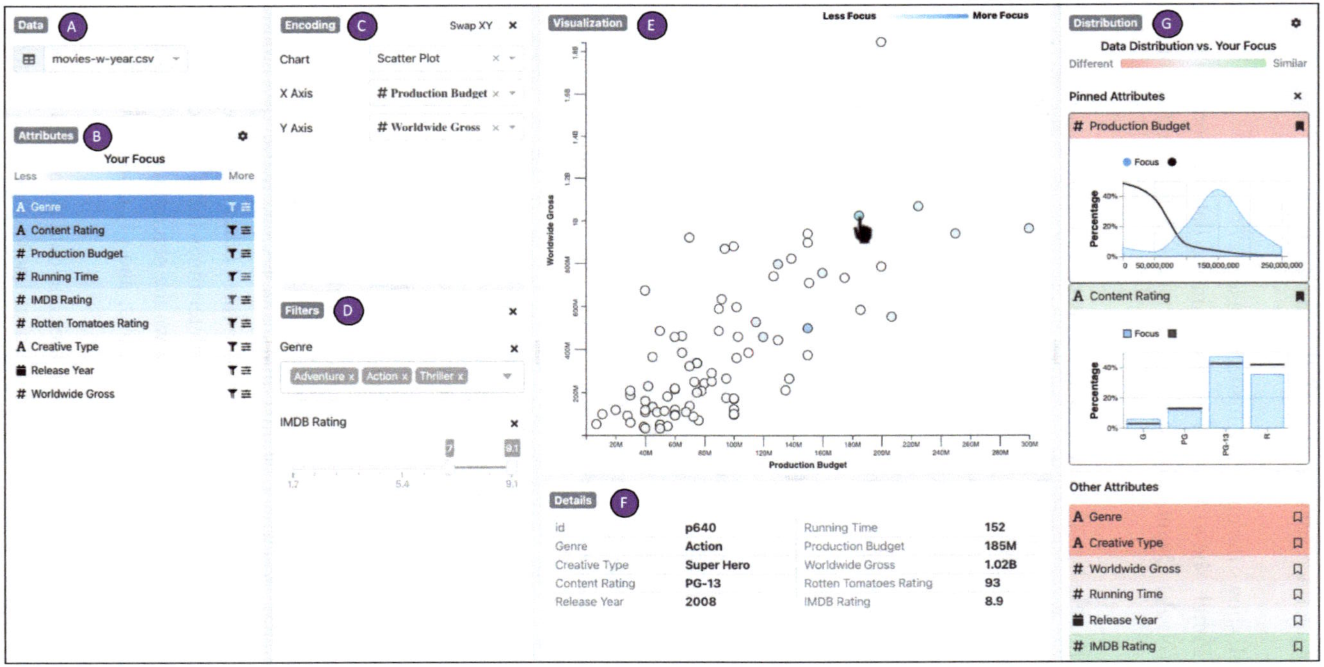

Fig. 7.3 The Lumos user interface includes traditional visual data analysis functions ((A) **Data Panel**, (B) **Attributes Panel**, (C) **Encoding Panel**, (D) **Filter Panel**), and shows analytic behavior as in-situ and ex-situ interaction traces in the (B) **Attributes Panel**, (E) **Visualization Canvas**, (F) **Details View**, and (G) **Distribution Panel** as the relation between a target distribution (e.g., the underlying data) and the user's analytic behavior

Details Release Year: 2008

id	Genre	Creative Type	Content Rating	Release Year	Running Time	Prod
p635	Action	Super Hero	PG-13	2008	126	
p637	Adventure	Super Hero	PG-13	2008	112	
p640	Action	Super Hero	PG-13	2008	152	
p659	Thriller	Contemporary Fiction	R	2008	129	
p670	Thriller	Dramatization	PG-13	2008	120	

Fig. 7.4 Details View (Aggregate). Hovering on an aggregate visual element in the *Visualization Canvas* (e.g., a bar in a bar chart) updates the Details View with a table of movies with *Release Year*=2008; darker blue=greater emphasis toward that data point

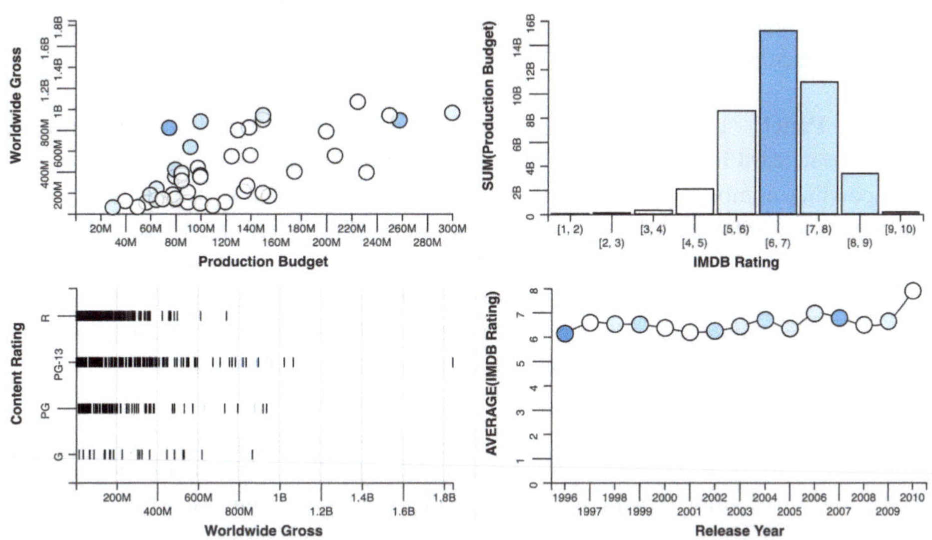

Fig. 7.5 Lumos captures mouseover interactions on visualization elements (e.g., points, bars) and leaves traces in-situ by coloring (fill) them in shades of blue in proportion to the relative frequency of interactions

In-situ Interaction Traces

Visualization Canvas. Lumos tracks user interactions with visual representations of data points (e.g., bars, lines, points, strips) and colors them on a white→blue scale based on the relative frequency of interactions, e.g., dark blue color represents more interactions (Fig. 7.5) and white represents no interactions **(DG2)**. Lumos captures mouseover interactions as a proxy for modeling analytic behavior from interactions with data points. Lumos employs a heuristic to ignore mouseovers that are active less than a 350 milliseconds threshold, regarded as random, accidental, or unintentional.

An interaction with a unit visualization (e.g., hovering on a point in a scatterplot of *Running Time* and *Worldwide Gross*) is handled differently than an interaction with an aggregate visualization (e.g., hovering on a bar showing average *Running Time* of *Action* movies). In the former scenario, Lumos treats it as one *complete* interaction with the corresponding data point incrementing its interaction counter by `1`. In the latter scenario, Lumos treats the interaction as a set of *partial* interactions with all constituent data points (e.g., all *Action* movies), incrementing their corresponding interaction counters by `1/N` where N=number of constituent data points.

Details View. When an aggregate visualization element (e.g., bar) is hovered on, the Details View below the chart shows a table with each data point. Lumos captures a mouseover on a table row, treats it as an interaction with the corresponding data point, and leaves an interaction trace by updating the table row's background color (Fig. 7.4) **(DG2)**.

Attribute Panel. Like data points, Lumos also tracks user interactions with attributes. Each attribute card in the Attribute Panel is colored on a white→blue scale based on the corresponding number of interactions (white=no interaction; darkest blue=most interactions). Lumos captures attribute assignments to encodings (e.g., X, Y) and filters (e.g., Gender=Male) as a proxy for modeling analytic behavior from interactions with attributes. These interactions are totaled and normalized relative to the most interacted attribute to determine the resultant shade of blue, e.g., in Fig. 7.3B, *Genre* has been interacted with most (dark blue) and *Worldwide Gross* has not been interacted with at all **(DG1)**.

Ex-situ Interaction Traces

Lumos allows users to compare their analytic behavior with a target distribution. Attribute cards in the Distribution Panel are colored on a red→grey→green scale (Fig. 7.3G) based on the difference between their respective analytic behavior and underlying distributions (as quantified by Wall et al.'s AD metric [121], described in Sect. 5.2) **(DG4)**. In this evaluation, we set the target distribution to the underlying data. A red background indicates that the user's analytic behavior is different from the target distribution, while a green background indicates similarity. For example, inspecting the visualization for *Production Budget* shows the analytic behavior peaking around `150M` (blue area) when most movies have a budget under `50M` (black curve); the magnitude of the deviation is high resulting in a red background. Similarly, the computed analytic behavior (blue bars) on *Content Rating* is more closely matching the underlying data (black strips) resulting in a green background.

Configuring Different Target Distributions

Users can define target interactive analytic behavior in multiple ways: (1) by **proportional** interactions across the various attribute distributions of the dataset, (2) by **equal** interactions across the categories of the dataset, and (3) by defining a **custom** target distribution of interactions across the data (**DG3**). For example, consider a dataset of job applicants, where 50% of applicants identify as male, 40% of applicants identify as female, and 10%

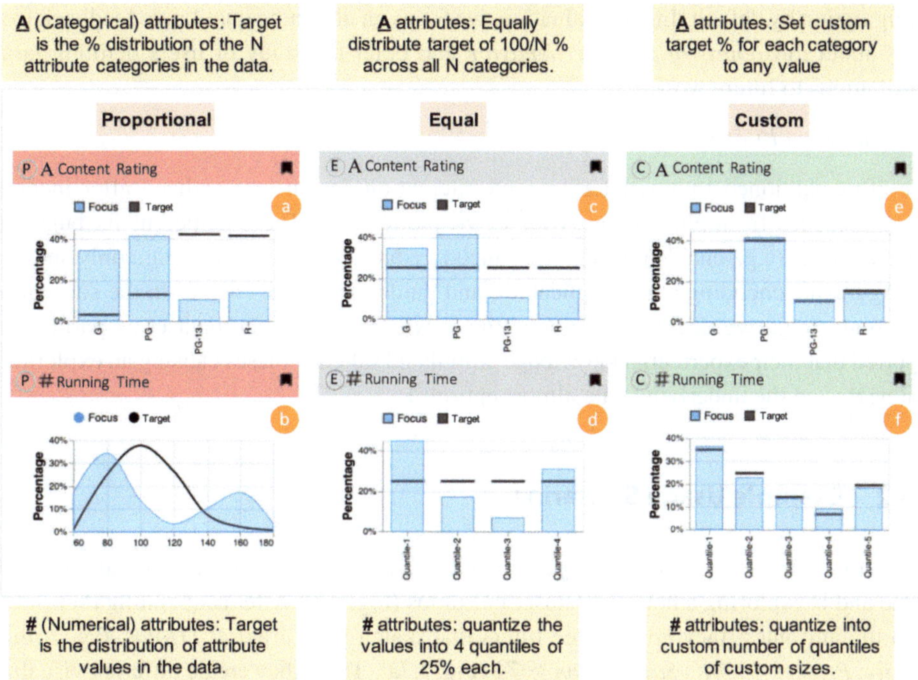

Fig. 7.6 Lumos supports ex-situ interaction traces for three modes of target distributions (*Proportional*, *Equal*, *Custom*). Lumos presents these targets in the charts as black curves/strips along with user behavior (blue area). Lumos also computes the difference between target and observed behavior and encodes it as the background color of the corresponding attribute card (red, grey, green colors where redder=more different; greener=more similar)

of applicants identify as non-binary. A **proportional** baseline would define the target distribution of interactive behavior such that 50% of interactions should be with male applicants, 40% with female applicants, and 10% with non-binary applicants, while an **equal** baseline would set the target distribution of interactions with 33.3% male applicants, 33.3% female applicants, and 33.3% non-binary applicants. If, for instance, diversity is a target in filling this particular role, then a **custom** baseline might be set, where the target interaction distribution is 40% female, 40% non-binary, and 20% male applicants. Figure 7.6 summarizes these settings in the context of a movies dataset for the Content Rating attribute, and shows (in blue) how the user's actual analytic behaviors compare to the target. Users can configure **proportional**, **equal**, or **custom** target distributions per attribute in Lumos. In the **custom** mode for categorical attributes, users are presented with an interactive bar chart where they can drag individual bars (each representing a category) to their desired relative weights. For quantitative attributes, users can sketch a target distribution by clicking (to add new quantiles) and dragging points in the presented interactive histogram. While Lumos supports all three target distribution variations, our study in the coming sections utilizes a **proportional** configuration by default. In this evaluation, we focus on obtaining qualitative feedback from users about *awareness of analytic behavior* and defer the study of users defining target distributions to future work.

Feedback from Pilot Studies

The aforementioned Lumos interface and interaction design was realized after incorporating feedback from pilot studies with `five` participants. These helped us fix bugs, add enhancements (e.g., enable resizing the panels), reduce errors and slips (e.g., only showing the Aggregate encoding when applicable), and guide the design choices in the subsequent user study (e.g., we decided to remove the movie *Title* from the dataset as participants reported that their experiential knowledge stimulated saliency biases causing an exploration that neglected the underlying data almost entirely).

7.4.2 Example Usage Scenarios

Scenario 1: Increasing awareness of analytic behavior Assume Austin is looking for a new home and is exploring a housing dataset in Lumos (Fig. 7.7). After acquainting themselves with the attributes, they apply three filters that match their criteria: {*Home Type=Single Family*; *Price* $\leq$ *\$300K*; *Satisfaction* $\geq$ *7*} (Fig. 7.7a). Then, they create different visualizations by specifying encodings (Chart Type, X, Y, and Aggregation) in the Encoding panel (Fig. 7.7b).

While interacting with these different visualizations, they observe visualization elements (e.g., bars, points) changing colors to different shades of blue. For example, in the scatterplot configuration with *Lot Area* (Y axis) and *Year* (X axis) (Fig. 7.7c), Austin observes their

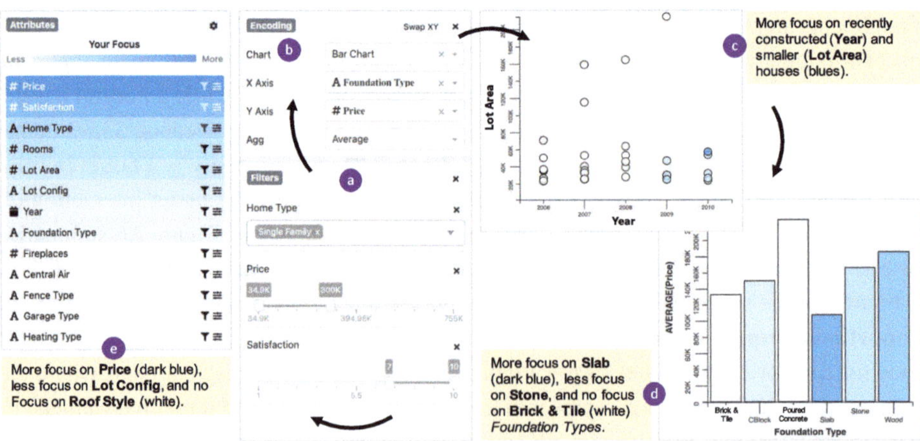

Fig. 7.7 Scenario 1: Lumos helping Austin be more aware of their data analyses

focus has been on smaller (*Lot Area* $\leq$ *60K*) and more recently constructed (*2009* $\leq$ *Year* $\leq$ *2010*) homes. Similarly, in the barchart configuration with *Foundation Type* (X axis) and *Average(Price)* (Y axis) (Fig. 7.7d), they observe they have not focused on `two` types of *Foundation Types*: {*Brick & Tile*, *Poured Concrete*} (white).

During their analyses, they also observe different shades of blue in the Attributes Panel (Fig. 7.7e) inferring they have not focused on all attributes equally, e.g., they focused more on *Price* and *Satisfaction* (darker blues), not so much on *Year* and *Foundation Type* (light blues) and not-at-all on *Fireplaces* and *Heating Type* (white).

After acknowledging that they did focus on the blue attributes, they start inspecting the `five` white attributes. They state they do not care about {*Lot Config* and *Fence Type*} but regret not focusing on the other `three` attributes (*Heating Type*, *Fireplaces*, *Central Air*) associated with climate control as the city faces severe winters. Accordingly, they apply new filters and encodings and continue their analyses.

In this way, Lumos helped Austin in house-hunting by making them more aware of their analytic behavior with focus on data and attributes.

Scenario 2: Mitigating biased analytic behavior

Taylor, a loan officer, is using Lumos to analyze loan applications to determine credit-worthiness. After exploring the dataset for a while, they observe a red *Home Ownership Type* attribute card and a green *Age* attribute card (Fig. 7.8a) in the Distribution Panel. They express happiness at not exhibiting any age bias but are concerned that their interactions with *Home Ownership Type* significantly deviate from the underlying data (target) distribution. They click on the card to toggle it open and begin inspecting the visualization. They observe they have unknowingly overemphasized on *Own* and *Rent* and underemphasized on *Mortgage Home Ownership Type*.

Willing to correct their behavior, they apply a (reverse) filter: {Home Ownership Type=*Mortgage*} (Fig. 7.8b) and analyze a few previously unconsidered (white) points (Fig. 7.8c). They finally see a greener *Home Ownership Type* card (Fig. 7.8d) and are more content.

7.4.3 Evaluation

Study Design

We conducted a between-subjects qualitative study with the aim of understanding how Lumos helps users increase awareness of their analytic behaviors. The study protocol was reviewed and approved by our institutional review board.

We recruited 24 participants (Gender: 14 male, 8 female, 1 other, 1 preferred not to say; Age in years: $\mu = 24.04$, $\sigma = 3.44$, median = 24, 1 preferred not to say). Participants included students, researchers, and industry professionals with diverse educational degrees (8 bachelors, 7 masters, 9 doctoral) all from a computing or related field (e.g., computer science, human-computer interaction, human-centered computing) with a self-reported

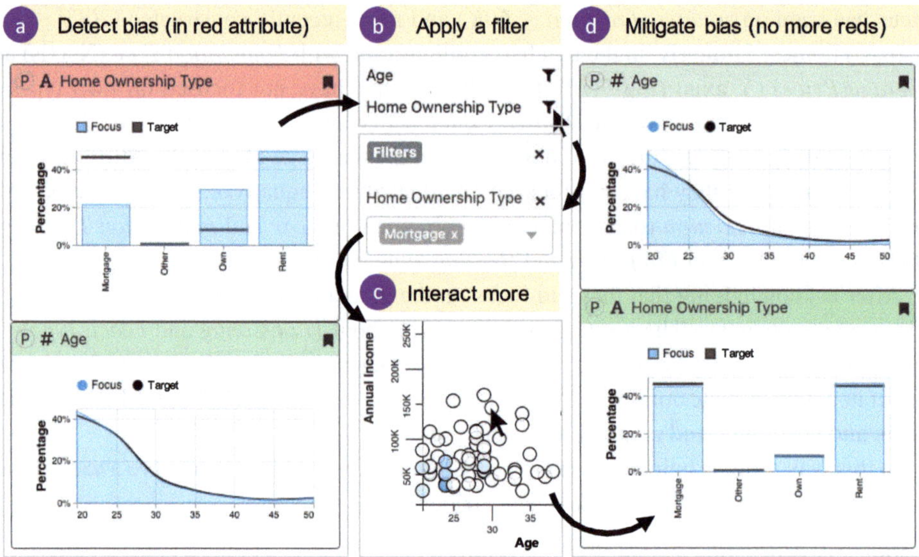

Fig. 7.8 Scenario 2: Lumos helping Taylor mitigate biased analytic behavior within the *Home Ownership Type* attribute categories

visualization literacy ≥ 3 (on a 5-point Likert scale). Sessions lasted approximately 1 h and participants were each compensated with a $15 Amazon gift card.

Due to the COVID-19 pandemic, we conducted the study remotely using online collaboration tools. The actual study was conducted over the Microsoft Teams (https://www.microsoft.com/en-us/microsoft-teams/group-chat-software) teleconferencing software. The experimenter provided participants access to the study environment by sharing their computer's screen and granting input control.

Participants were randomly divided into either a **Control** or **Awareness** condition, which determined the version of the system they used for the study. Both conditions had roughly equal numbers of participants with bachelors, masters, and doctoral degrees (Awareness: bachelors=4, masters=4, doctoral=4; Control: bachelors=4, masters=3, doctoral=5). Participants in the Control condition *did not* see the Distribution Panel (along with the ex-situ interaction traces) nor did they see the in-situ interaction traces in the Visualization Canvas, Details View, and Attribute Panel. We pre-configured the target distribution to *Proportional* and hid the Settings Panel for both conditions.

Each participant first electronically signed a consent form. Then, depending on the study condition, they saw a 2 min (Control) or 4 min (Awareness) video tutorial that demonstrated the features of Lumos. Participants were then asked to perform a practice task using a dataset of cars to interact with and get acquainted with Lumos. After practice, we tasked participants to *"analyze a dataset of movies to recommend the characteristics of movies that a movie production company should make next"*. The dataset consisted of `709` movies (rows) and 9 attributes (columns): *Production Budget* (**#**), *Worldwide Gross* (**#**), *Running Time* (**#**), *IMDB Rating* (**#**), *Rotten Tomatoes Rating* (**#**), *Release Year* (📅), *Content Rating* (**A**), *Genre* (**A**), and *Creative Type* (**A**). Participants were encouraged to think aloud during the task. Throughout the task, their interactions were recorded. The study ended with a debriefing in which participants discussed their overall experience with the system, provided suggestions for improvements, and completed a questionnaire to rate the usefulness of Lumos features. All sessions were screen- and audio-recorded for qualitative analysis.

Hypotheses

We structure our study analysis according to the following hypotheses:

H1 Interaction traces will increase awareness of analytic behavior.
H2 There will be differences in interactive behaviors of Awareness versus Control participants (as measured by bias metrics [121], differences in use of filters, and number of charts created).
H3 Participants will find the ex-situ awareness features to have greater utility than in-situ awareness features.
H4 Participants in the Awareness condition will react to interaction traces in ways to reduce potential biased analytic behaviors.

Results

Below, we present findings from the study and discuss them in context of participant feedback. P_A01-P_A12 and P_C13-P_C24 refer to the 24 participants in Awareness and Control conditions, respectively. Participant quotes and moments of awareness were coded and categorized using affinity diagramming. One administrator came up with an initial set of categories that were then refined during iterations with three other administrators until a consensus was reached.

General Feedback

Overall participant feedback was positive with P_A01 commenting, *"[they] haven't seen many things like [Lumos] before...really good technique"*. P_A09 mentioned that *"[Lumos] can remove the internal bias of things users think are of the most interest"*. Participants found *"the ability to keep track of [their] provenance, interaction history [as] interesting"* (P_A08) and *"communicating it back [how they are doing] as something [they] would use in [their] tools"* (P_A07). P_A09 found the Distribution panel *"a great idea to show users what their focus was"* and P_A05 found it *"very helpful as [they] don't need to create visualizations in the Vis panel for each attribute to see [their] distributions"*. P_A05 suggested *"integrating this tool into existing tools such as Tableau [as] they don't have a feature that tells [them] what attributes haven't been used yet"* (P_A12). P_A07 suggested *"there are lots of use-cases for this technique in journalism and social media, e.g., you have only looked at Trump's negative tweets, but what about Biden's?"*. Two participants found the Distribution Panel less useful as they *"didn't know exactly what to do about the [red-green] cards"* (P_A5) or felt it *"out of focus"* on the right side of the screen (P_A10).

Usefulness Scores: Lumos user interface

Figure 7.9a summarizes Lumos's usefulness scores (*1=Not useful at all; 5=Extremely useful*) as reported in the post-study questionnaire:

Attribute Panel. Participants generally found the attribute list useful ($\geq$3 out of 5, median$_A$=4, median$_C$=4.5) along with *"their data types"* (P_C21) *"unlike, e.g., Excel where they aren't always on-screen"* (P_C17).

Encoding Panel. 23 participants found the Encoding Panel to be useful (median$_A$=4, median$_C$=4, *"it is standard in a good way"*-P_C21) except P_C23 who found it *"only slightly useful"*. Four participants noted that the system messages to fix incorrect encodings were intuitive and helpful (*"it is sometimes hard to know what's wrong in Tableau"*-P_A05) but two found them confusing and suggested the app *"prevent [them] from choosing incorrect encodings"* (P_C18) by *"filtering out the chart types that are not allowed"* (P_A11). Five participants also suggested additional features (*"add color as a third encoding"* - P_C24) and enhancements (*"support drag-drop attributes to the Encoding Panel"* - P_C{17,18,19}, *"support text entry for the dropdowns"* - P_A12).

Filter Panel. Participants utilized filters (median$_A$=4, median$_C$=4.5) *"to remove outliers and to confirm hypotheses about the data"* (P_C22), to see the different values for a categorical attribute (P_C19, P_C23), and to mitigate any unconscious biased analytic behavior (e.g., *"Comedy and Drama are high percentage in the dataset, and I haven't interacted with them at all, so it might be worth my time to look at them."*-P_A09). P_A10 did not utilize filters as they were *"being more exploratory with [their] analysis and if [they] wanted to look at finer details, [they] would have used more filters."* Three participants also requested enhancements to *"specify precise inputs [for Q,T values]"* (P_C22), *"allow hover on a value in the categorical filter and highlight in the Visualization"* (P_A09), and *"support select- and deselect- all for categorical values"* (P_A11).

Visualization Canvas. 23 participants found the Visualization Canvas useful (median$_A$=5, median$_C$=4.5), utilizing it to *"observe patterns and outliers"* (P_A10), and *"see the different categories and values in the attributes"* (P_C14). P_C18 *"did not find it useful because a third attribute encoding, e.g., color was not supported"*.

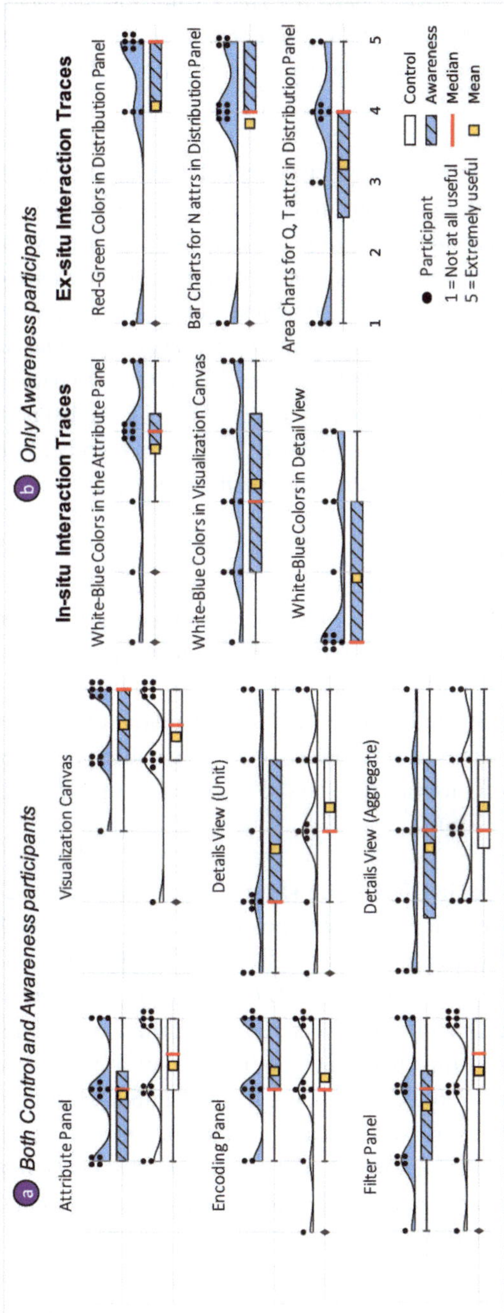

Fig. 7.9 Summary of usefulness scores of all Lumos features as reported by participants in the post-study questionnaire, as RainCloudPlots [8]

Details View (Unit). This view received mixed usefulness scores from our participants ($median_A$=2, $median_C$=3). P_C17 *"liked the Details portion and being able to hover over points for more details."* P_C19 and P_A11 *"didn't find the Details view for single data points super useful"* because they wanted the name of the film to bring prior knowledge to the analysis and spark different hypotheses.[1]

Details View (Agg). This view also received mixed usefulness scores from our participants ($median_A$=3, $median_C$=3). P_C18 found it to be *"really useful because it is not apparent from the bar shape and size that some bars only have one point in them versus some bars having six or seven points"*. P_A07 said *"[they] don't really hover on things in e.g., a scatterplot but information shown on hovering a bar chart [the details view agg] was awesome because you showed individual data"*. Also, *"it shows all information in one space."* (P_C20, P_C24) and *"could be useful for multivariate hypotheses"* (P_A10). Two participants found it *"hard to draw conclusions from lists of words and data"* (P_C17, P_C19) and preferred to see the information visually utilizing the Details View *"only as a reference"* (P_C19). P_C18 utilized the Details View *"because the visualization wasn't as helpful"*.

Usefulness Scores: Lumos technique

Participants in the Awareness condition also saw in-situ and ex-situ (Distribution Panel) interaction traces in the user interface. Figure 7.9b summarizes the usefulness scores (*1=Not useful at all; 5=Extremely useful*) as reported in the post-study questionnaire:

Difference between analytic behavior and target distribution. Ten participants found the difference between their analytic behavior and the underlying data (red-green coloring in Distribution Panel) very useful ($median_A$=5) *"giving a sense if I'm looking at the dataset in an unbiased way" (P_A06).*

Ex-situ interaction traces in Distribution Panel. The overall feedback on the ex-situ interaction traces in the Distribution Panel was positive; participants found the bar charts for N attributes ($median_A$=4) more useful than the area charts for Q,T attributes ($median_A$=4) with P_A06 nicely summarizing, *"I can track and channel my focus based on discrete bar charts by applying a filter...but it is difficult to discretize and track a continuous [Q,T] attribute"*. For eight users, these real-time traces helped increase awareness (*"Geez, I haven't looked at Drama movies at all"*-P_A07), influencing them to interact differently (e.g., by creating a bar chart with *Genre* to inspect movies of other potentially underemphasized genres) while two participants either ignored them (*"I never looked at the individual distributions of attributes"*-P_A12) or preferred to look at them after analysis *"as the bars will be moving, and that's distracting"* (P_A09).

[1] Note: the movie *Title* was deliberately not shown in the Detail View to prevent personal experiences with data to influence the analysis.

In-situ interaction traces in Visualization Canvas. There was mixed response to the in-situ interaction traces (white-blue colors) in the Visualization Canvas (median_A=3). For P_A06, they *"helped in tracking visited points"* nudging them to interact with points that were not interacted yet, while for P_A05 they were confusing and distracting, nudging them to re-interact with them. P_A08 did not want the colors to stay persistent but *"be able to clear existing interactions and start a new session with a new set of model movies for comparison"*.

In-situ interaction traces in Details View. Only four participants found the in-situ interaction traces (white-blue colors) in the Details View to be useful (median_A=1).

In-situ interaction traces in Attribute Panel. Participants generally found the in-situ interaction traces (white-blue colors) in the Attribute Panel to be useful (median_A=4). For P_A05, they helped increase awareness of already-interacted attributes (*"I see that I have spent a lot of time on Release Year so I'll now see something else"*) while for P_A12, they required some time to get acquainted with (*"the coloring in the Attributes panel...I didn't use it initially, and later on it hit me that I had this feature. Once I noticed it, it was very useful"*).

Summary. Comparing the distributions of scores for the aforementioned features (Fig. 7.9b), participants found the ex-situ interaction traces more useful than the in-situ interaction traces, supporting **H3**, consistent with experimental results from [119]. We believe this is because in-situ interaction traces are always visible to a user whereas ex-situ interaction traces can be used more on-demand without side-tracking the analysis task at hand. Furthermore, in-situ traces block an otherwise common attribute encoding channel, *color*, that can be undesirable for and cause inconvenience to some users.

Awareness Moments

In-situ traces in the Attributes Panel. There were instances when Control participants expressed a need for tracking the already-interacted attributes. For example, we observed P_C13 use hand gestures to recollect and count the attributes that they had already visited and P_C14 exclaimed, *"I hope I have interacted with all (attributes)"*. Awareness participants, on the other hand, saw the interaction traces and had several instances of *awareness* during their respective analyses. Two participants *acknowledged* their choices (*"I don't think Release Year should matter too much, hence I am not interacting with it."*-P_A04, *"I don't think Runnning Time is important to me"*-P_A07) while two participants also suggested correcting future course via *interaction* (*"I see that I have spent a lot of time on Release Year so I'll now see something else"*-P_A05, *"[on seeing a white attribute bar] now I'm going to interact with Running Time"*-P_A22), also supporting **H4**. Two participants reflected upon their choices after the study while they were answering questions pertaining to self-reported focus on individual attributes in the post-study questionnaire (*"actually I forgot, had I remembered,* *it might have been interesting to not click on the same thing over and over."*-P_A03, *"I didn't use the blue attributes panel but now that I see these questions, I would've seen them more"*-P_A08). These and the desire for awareness moments by Control participants validate **H1**.

In-situ traces in the Visualization Canvas and Details View. Eight participants found the in-situ interaction traces in the visualization canvas to be useful; two participants took some time to get acquainted with them (*"very useful but I learnt about them slightly afterwards"*-P_A01, *"I was initially confused but then over use I got used to them and found them useful in tracking visited points"*-P_A06). P_A03 found the colors to be useful but questioned the technique because *"if it is based on [me] hovering on a point again and again, it might not be 100% correct."* P_A05 was distracted and *"getting drawn to the visited points (instead of the white un-visited points)."* There was minimal commenting on the in-situ traces in the Details View but it led to some awareness for P_A04 who hovered on an uninteracted (white) bar in a bar chart and observed *"there aren't many blue rows which means I haven't been focusing on it."*

Ex-situ traces in the Distribution Panel. There were multiple instances of awareness among participants (supporting **H4**). P_A04 verified the interactions traces by comparing it with ground truth (*"distribution of my focus on Running Time (blue) is representative of the applied filter"*). P_A05 reflected upon seeing three red attributes in the Distribution Panel and hypothesized that they were *"just thinking aloud and exploring and will (now) follow a more targeted approach"*. P_A05 reflected upon seeing a red *Content Rating* attribute (*"Seems I didn't interact with R-rated movies enough so this view nudges me toward those"*) and applied a filter to show only R-rated movies. P_A07 reflected upon the white *Drama* category in the Distribution Panel and justified that they *"didn't care about Dramatization movies [...] who cares?"* At one point, P_A07 observed many red cards and exclaimed, *"this is so biased but whatever."* P_A11 tried to correlate the effects of their interactions with different attributes (*"I noticed that Adventure is representative of most values in Rotten Tomatoes Rating and IMDB Rating [...] This is because I mostly interacted with just Adventure movies and that caused those attributes [in the Distribution Panel] to be colored green"*). P_A10 did not use the Distribution Panel as they *"didn't know how to use it in the context of what [they were] doing"*.

Interactive Behaviors Across Experimental Conditions

Overall, Awareness participants exhibited more diverse behavior in terms of interactions with data points and attributes than Control (Fig. 7.10a; μ_A=256, σ_A=115; μ_C=289, σ_C=66), validating **H2**.

Interactions with Data Points. Figure 7.10b shows the distribution of all interactions with data points across conditions (μ_A=89.92, σ_A=43.83, μ_C=98.75; σ_C=35.67). P_A08, on seeing the interaction traces, became fearful of skewing their interactions. Both conditions

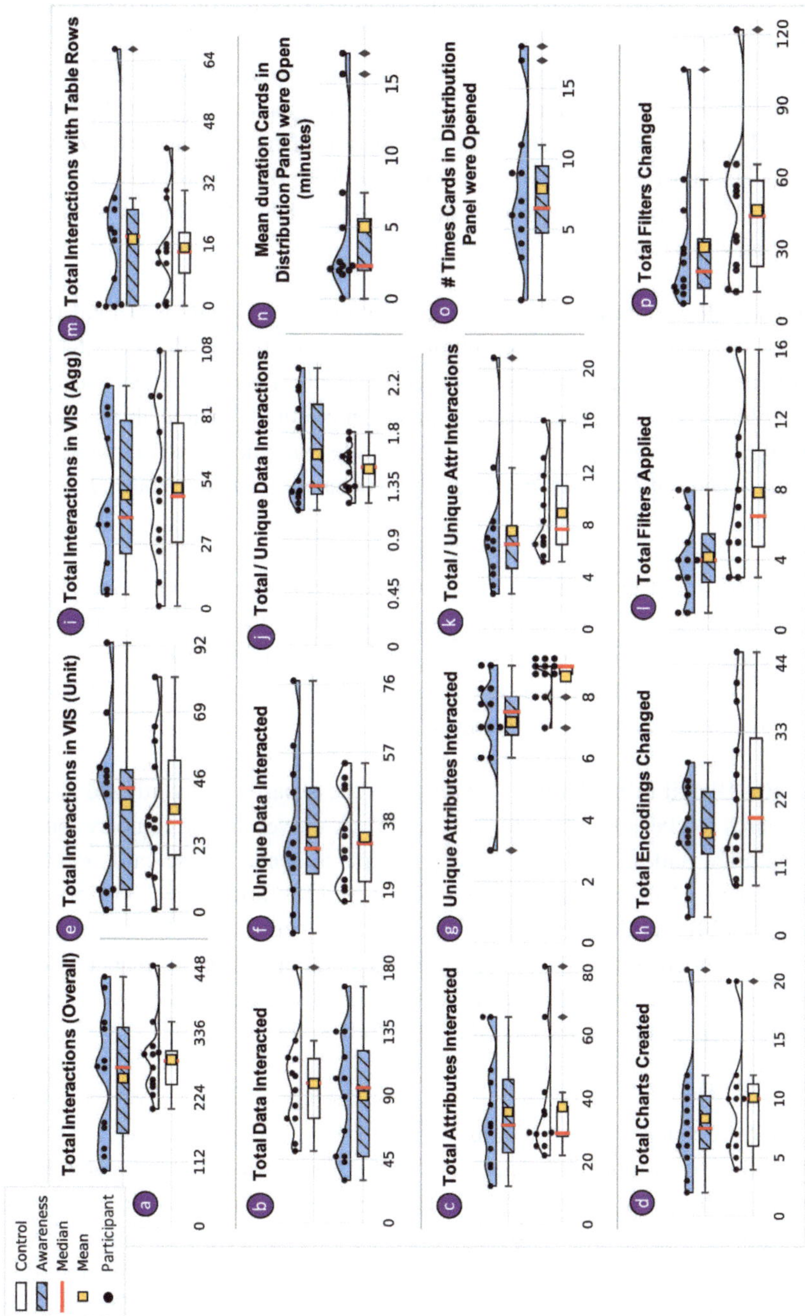

Fig. 7.10 Summary of quantitative findings across the two experimental conditions, as RainCloudPlots [8]. VIS=Visualization; Attr=Attribute

interacted with a similar number of unique data points (Fig. 7.10f; μ_A=35.08, σ_A=20.18, μ_C=33.67; σ_C=13.36) indicating that the in-situ blue colorings did not nudge participants to interact with more white points (in fact, P_A05 felt drawn toward the blue points). Interestingly, this result seems different from the one found by Hindsight [42], where participants tended to interact with more unique points when data were colored based on prior interactions. Figures 7.10(e,i,m) show the distribution of interactions with (e) data points in unit visualizations (μ_A=37.33, σ_A=27.93, μ_C=35.50; σ_C=23.68), (i) aggregate visualizations (μ_A=47.3, σ_A=32.53, μ_C=50.5; σ_C=33.51), and (m) the Details View table rows (μ_A=17.33, σ_A=19.11, μ_C=15.08; σ_C=12.62).

Interactions with Attributes. Figure 7.10c, d show the distribution of total (μ_A=35.75, σ_A=17.8, μ_C=37.42; σ_C=18.25) and unique interactions (μ_A=7.17, σ_A=1.64, μ_C=8.67, σ_C=0.65) with attributes across conditions, respectively. Four awareness participants did not interact with certain attributes as they were not important to their analysis (*"I don't think this [Running Time] is important to me"*-P_A11) while two others tried to interact with more attributes to try and mitigate their biased analytic behaviors with attributes (P_A05) and data points (P_A07).

Charts, Filters, and Encodings. On average, Control participants created more charts (Fig. 7.10d; μ_A=8.33, σ_A=5.02, μ_C=10.08, σ_C=5.28), changed more encodings (Fig. 7.10h; μ_A=16.58, σ_A=7.86, μ_C=23.00, σ_C=13.11), and applied (Fig. 7.10l; μ_A=4.17, σ_A=2.44, μ_C=7.83, σ_C=4.57)) and changed (Fig. 7.10p; μ_A=31.58, σ_A=28.05, μ_C=46.83, σ_C=30.50)) more filters than the Awareness participants.

Cards in the Distribution Panel. Eleven awareness participants utilized the Distribution Panel cards to check the data distribution (black curve) and/or compare their analytic behavior with the underlying data. These participants opened these cards multiple times at different points of their analysis (Fig. 7.10o, μ_A=7.91 times, σ_A=5.35 times) and kept them open for different durations (Fig. 7.10n, μ_A=5.00 min, σ_A=5.65 min). P_A09 opened the cards but commented that they would prefer analyzing their performance *after* the task was over. One awareness participant did not utilize these cards as they were content with the red-green backgrounds of the closed cards (P_A12).

Temporal Analysis

To understand the cadence of interactions, we plotted hover, filter, encoding, and Distribution Panel interactions over time, and superimposed the AD metric [121] (representing analytic behavior) for each attribute. We found that for some participants who did not interact with these UI components, their analytic behavior was skewed for specific attributes (as evidenced by high AD metric values for *Genre* for P_C16 in Fig. 7.11a). For others, interactions led to significant shifts in analytic behavior. For instance, while P_C13 used *Worldwide Gross* as an encoding in a chart, their analytic behavior toward the attribute became more proportional to the underlying data (Fig. 7.11b). While a filter was applied to IMDB Rating ($\geq$2.3 and

≤9.1), P_A09's analytic behavior began to deviate from the underlying data as shown by increasing AD metric value (Fig. 7.11c). Lastly, while we found instances where inspecting the Distribution Panel and accordingly interacting with the visualization appear to reduce biased analytic behavior (e.g., Fig. 7.11d, P_A01), such instances were not abundant among our participants. Thus, temporal analysis provides minimal evidence to support **H4**, but interaction traces provided participants opportunities to reflect on their analytic behavior, judging how their process differs from the baseline, and if that difference is appropriate for their task.

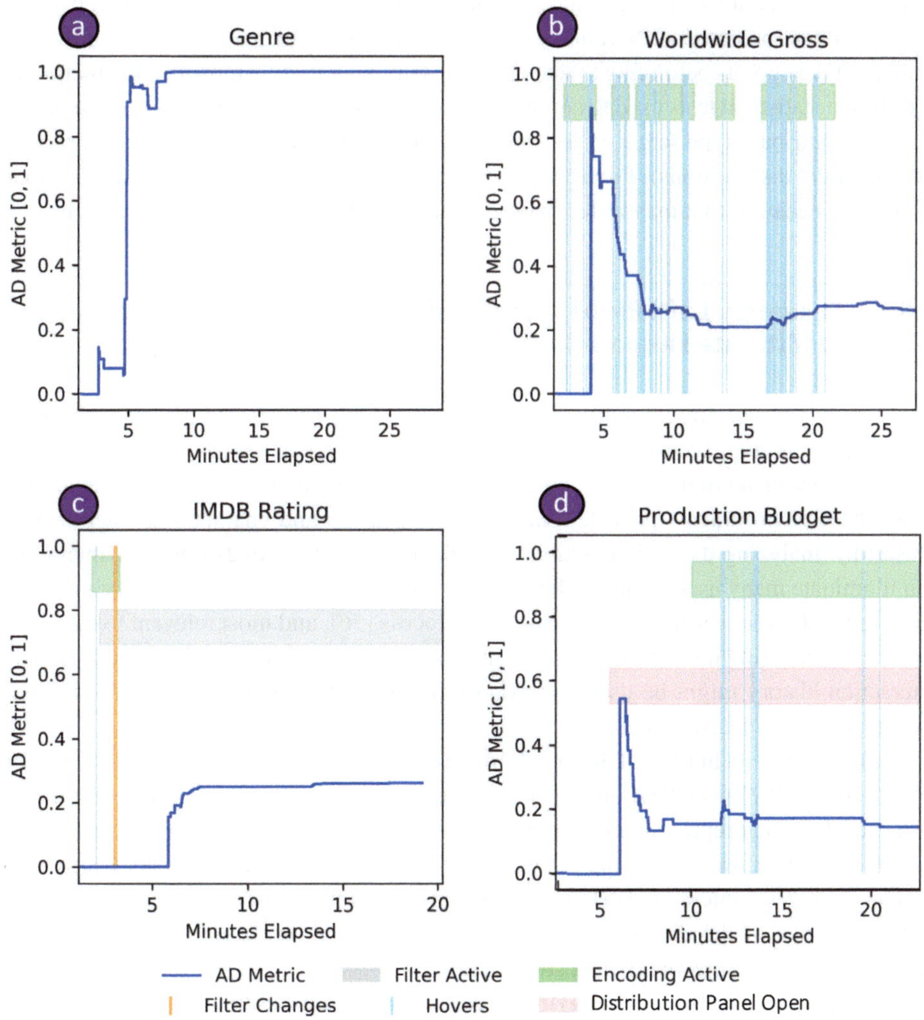

Fig. 7.11 AD metric values over time for **a** P_C16, **b** P_C13, **c** P_A09, and **d** P_A01 for specific attributes

7.4.4 Limitations

Lumos currently supports only a small set of visualization types; however, we chose them to test across different aggregation types. Also, analytic behavior is modeled only from interactions, which may not be a complete proxy for attention; in the future, one may consider user gaze or other sources to more accurately approximate it. Next, Lumos models analytic behavior by equally weighting the interactions. As P_A06 suggested to *"remove older interactions, say only keep the most recent 100 or 200 of them [as] people lose attention [over time]"*, we may consider interaction recency in the future.

Finally, interactions with aggregate visualizations (e.g., hovering on a bar in a bar chart) are currently considered as N equally weighted interactions of magnitude 1/N where N = number of data points belonging to that element. This has variable impact on the metrics due to different statistical tests used to compute the analytic behavior model (AD [121]) depending on the attribute type (e.g., χ^2 test for categorical attributes, Kolmogorov-Smirnov test for numerical distributions). Future work can explore alternative computations for models of analytic behavior that may reflect a user's attention and intentions more precisely.

7.5 Research: Left, Right, and Gender—Exploring Interaction Traces to Mitigate Human Biases

As the sheer volume and ubiquity of data increases, data analysis and decision making are increasingly taking place within digital environments, where humans and machines collaborate and coordinate to inform outcomes, facilitated by interactive visual representations of data. These environments provide a new way to measure and characterize cognitive processes: by analyzing users' interactions with data during use. Analyzing user interactions can illuminate many aspects about the user and their process, including identifying personality traits [20], recovering a user's reasoning process [36], and most relevant to the present work, characterizing human biases [121]. In this work, we explore how showing a user prior interaction history might be used to **mitigate potential biases** that may be driving one's data analysis and decision making.

We utilize the technique of **interaction traces** defined in the previous section, a form of provenance [87] visualization in which a user's own previous interactions with the data influence the visual representations in the interface. We show interaction traces in two ways: in-situ interaction traces alter the color of visited data points in a scatterplot based on the frequency of prior interactions (Fig. 7.12E), and ex-situ interaction traces are shown in an additional view of the data that compares the distribution of a user's interactions to the underlying distributions in the data (Fig. 7.12F).

We operationalize **biased behavior** as deviation from a baseline of equally probable interactions with any data point. It can be conceptualized as a model mechanism [118], captured using bias metrics [121], and may correspond to other notions of societal or cognitive

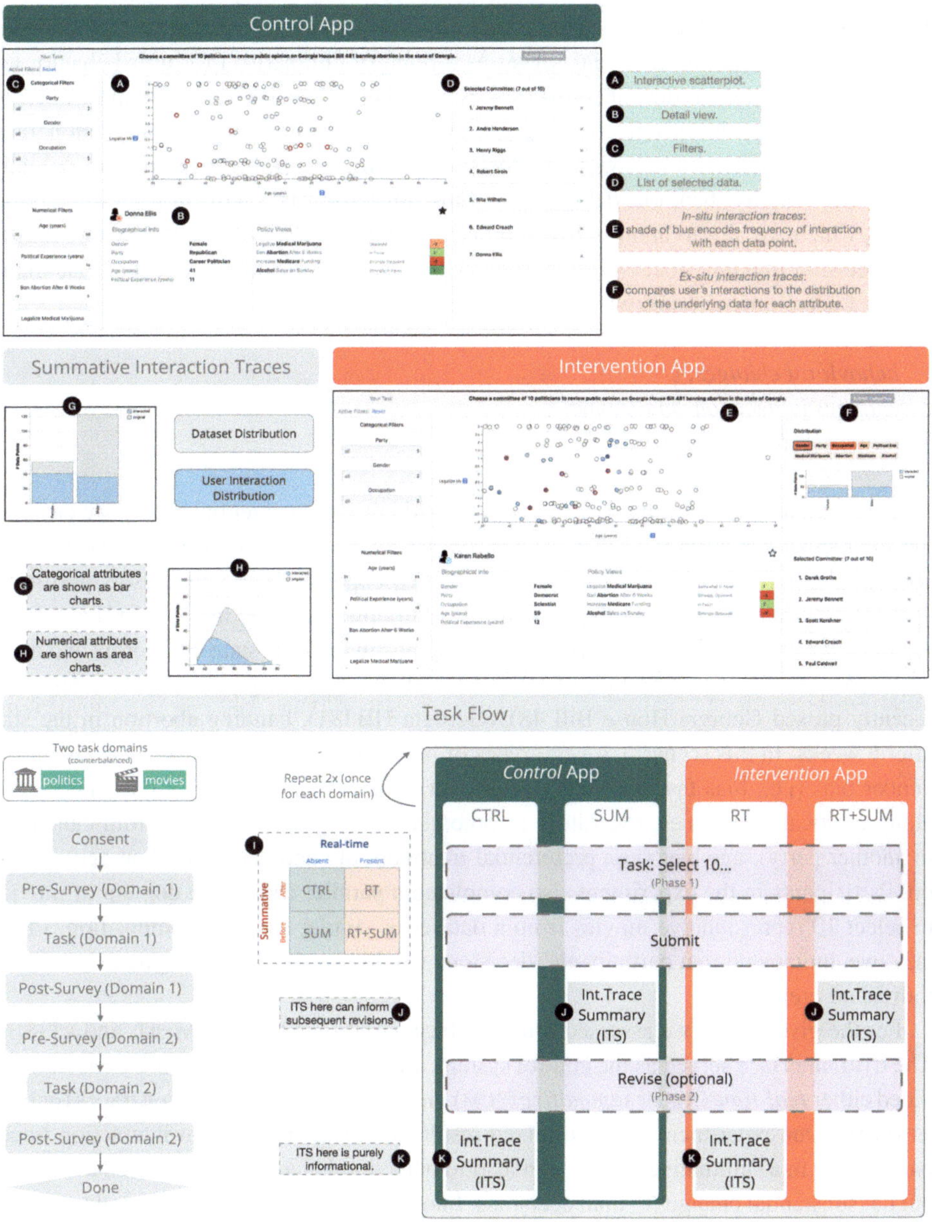

Fig. 7.12 A summary of the interfaces and procedure for this experiment

bias. Similarly then, a **biased decision** is one which reflects choices that are not proportional to the data. This definition of bias serves as a point of comparison for user behavior and decision making, but, as described in [121], is not inherently negative and requires interpretation in context by the user given their goals. We posit that visualization of interaction traces will lead to reflection on behavior and decision making, increasing awareness of potential biases. Importantly then, our definition of **bias mitigation** is a reduction in *unconscious* biases, which we aim to address by promoting user reflection [100] about factors driving their decision-making processes. In particular, we examine the effectiveness of visualizing traces of users' interactions, where effectiveness is measured by

1. *behavioral changes*,
2. *changes in decisions made*, and
3. *increased cognitive awareness.*

To assess the impact of interaction traces toward mitigating potential biases, we designed the interactive scatterplot-based visualization system, Lumos, described in the previous section (Fig. 7.12). We conducted a crowd-sourced experiment in which users performed two decision-making tasks in the domains of (1) politics and (2) movies. In the political scenario, we curated a dataset of fictitious politicians in the U.S. state of Georgia and asked participants to select a committee of 10 responsible for reviewing public opinion about the recently passed Georgia House Bill 481 (Georgia HB481), banning abortion in the state after 6 weeks. In this scenario, several types of bias may have impacted analysis, including gender bias (i.e., bias favoring one gender over another), political party bias (i.e., voting along political party lines, regardless of potential ideological alignment from candidates in another party), age bias (i.e., preferential treatment of candidates based on age), and so on. Participants in the experiment also completed a parallel task in the domain of movies: to select 10 representative movies from a dataset of similar size and composition. In this task, we anticipated that participants' decisions would be driven by idiosyncrasies of their individual preferences.

For the given tasks, we assessed four interface variations: CTRL, SUM, RT, and RT+SUM. The CTRL interface served as the control system, which we compared to variations that provided either *real-time* (RT) or *summative* (SUM) views of the user's interaction traces (or both, RT+SUM). Our experiments yielded mixed results, offering support that interaction traces, particularly in a summative format, can lead to behavioral changes or increased awareness, but not substantial changes to final decisions. Interestingly, we find that increased awareness of unconscious biases may lead to amplification of individuals' conscious, intentional biases. We emphasize that regardless of domain, our goal is not to address overt biases (e.g., in the form of discrimination) in this work; rather, we believe visualization *can* have an impact on increasing user awareness of potential unconscious biases that may impact decision making in critical ways.

This work was originally published at IEEE VIS in 2021 [119]. We highlight the following contributions:

1. we utilize a technique for showing interaction history, (referred to as *interaction traces*, Sect. 7.5.1),
2. we present results of three formative in-lab studies that describe exploratory and qualitative findings (Sect. 7.5.2), and
3. we present results of a crowd-sourced study that describes quantitative effects of interaction traces (Sect. 7.5.3).

In the following sections, we present a description of the datasets and interface used in the studies, findings from the in-lab and crowd-sourced experiments, and a discussion of how these results can inform the design of future systems that can mitigate potentially biased analyses.

7.5.1 Methodology

To study the effect of visualizing interaction traces toward mitigating bias, we conducted a series of in-lab studies and a crowd-sourced experiment to test four interface variations (CTRL, SUM, RT, RT+SUM). In this section, we describe two tasks and datasets in the domains of politics and movies and the implementation of a visualization system that realizes interaction traces to serve as the testbed for subsequent experiments.

Tasks and Datasets

We selected two complementary tasks (counterbalanced within subjects) to observe how people would respond to interaction traces in the presence of a variety of potential biases, described below for each task.

Politics

Task. The USA has a two-party political system: Democrats and Republicans [7]. In Georgia's General Assembly, committees may be formed to explore complex issues, draft legislation, and make recommendations [48]. Many such committees, particularly subcommittees focused on specific issues, may be formed by top-down appointment [48]. With membership in committees often decided by an individual or by few, the decision can be subject to an individual's biases.

In May 2019, Georgia's incumbent Governor Brian Kemp signed Georgia House Bill 481 (Georgia HB481) banning abortion after 6 weeks (earlier than the previous state law of 20 weeks) [95]. Scheduled to take effect in January 2020, the bill was received by the public with significant controversy.[2] Supporters of Georgia HB481 (colloquially referred to as a "Heartbeat Bill") hoped it would lead to overturning of Roe v. Wade, 410 U.S. 113 (US federal court decision protecting a woman's right to an abortion, 1973), while opponents hoped to challenge the bill before it became law.

Given a dataset of fictitious politicians, participants were given the following task:

> Imagine you are engaged in political decision-making in the state of Georgia. The debate about abortion is ongoing, with variations of these bills cropping up across other states in the United States, which can potentially learn from the ongoing debate in Georgia. Select a committee of 10 candidates that you feel should review public opinion in Georgia on the controversial Georgia HB481.

We selected this task to simulate a realistic decision-making scenario in American politics and evaluate our interventions in a politically and socially relevant context. Furthermore, this topic and dataset can elicit a number of factors that may influence an individual's decision-making process, including personal preferences as well as multiple types of social biases (e.g., gender bias or political party bias), both conscious and unconscious.

Dataset.

We generated a dataset of 180 fictitious politicians, representing the composition of the Georgia General Assembly [47]. Each row in the dataset represents a politician, described by the following attributes: *Gender*, *Political Party*, *Occupation*, *Age*, and *Experience*, along with numerical representations $\in [-3, 3]$ of the politician's view on topics such as *Banning Abortion After 6 Weeks*, *Legalizing Medical Marijuana*, *Increasing Medicare Funding*, and *Banning Alcohol Sales on Sundays* (positive numbers indicate that the politician is in favor, while negative numbers indicate that the politician is opposed). Politicians' names are artificially generated from US census data [93]. The dataset contains 59% Republicans, of which 14% are female; and 41% Democrats, of which 57% are female, mimicking the distributions in the Georgia General Assembly [126]. The ages, political experience, and occupations were derived from data on the 115th U.S. House of Representatives [81]. The policy views were generated to represent general party voting trends (e.g., Democrats tend to be opposed to banning abortion, while Republicans tend to be in favor of the ban) with the strength of those views representing recent increasing polarity [94] in the USA political system (e.g., fewer politicians have neutral positions or positions against the party trend).

Movies

Task. Given a dataset of fictitious movies, participants were given the following task:

> Analyze the data to pick 10 movies that you feel represent the collection of movies in the dataset as a whole.

[2] A federal judge permanently blocked Georgia HB481 in July 2020, finding it in violation of the U.S. Constitution [91].

We selected this task to complement the political scenario. It represents a parallel task (selecting a representative subset) in a domain that the general public is familiar with (movies). We hypothesize that this task may elicit an entirely different set of (less obviously dangerous) biases, based on idiosyncrasies in one's movie preferences. For instance, participants may make selections for movies by focusing on attributes of the data that are most familiar to them (e.g., *Rotten Tomatoes Rating*) while disregarding others that have a lesser impact on their own movie habits (e.g., *Running Time*). The instructions for both tasks were intentionally vague to avoid suggesting any particular criteria for selecting politicians/movies.

Dataset. The movies dataset was adapted [1] to match the general structure of the political dataset. We sampled 180 movies from the dataset and selected 9 attributes in total (3 categorical, 6 numerical): *Content Rating*, *Genre*, *Creative Type*, *Worldwide Gross*, *Production Budget*, *Release Year*, *Running Time*, *Rotten Tomatoes Rating*, and *IMDB Rating* to match the dimensionality of the political dataset. In pilot studies, we found that (1) real movie titles were problematic because participants relied heavily on familiarity of titles rather than the data before them; and (2) anonymized identifiers (e.g., "Movie1", "Movie2", ...) led participants to be less engaged with the task. For consistency with the political scenario, we generated fictitious movie titles[3] so that participants would be more engaged with the task while not relying only on familiarity of titles. Complete datasets and analyses are included in supplemental materials.[4]

System

Overview. For our experiments, we utilized a simplified version of Lumos [84] (Sect. 7.4), a visualization system to support data exploration while promoting reflection and awareness during visual data analysis. To assess the effectiveness of visualizing interaction traces, we produced two versions of the visualization system: a Control version of the interface, and an Intervention version of the interface, which was modified to visualize traces of the user's interactions with the data in real-time (Fig. 7.12). Components A-D in Fig. 7.12 are common across the Control and Intervention interfaces. The primary view is an interactive scatterplot (A), where the x- and y-axes can be set to represent attributes of the data via selection in a drop-down menu. Hovering on a point (politician/movie) in the scatterplot populates the detail view (B), which shows all of the attributes of that data point. Filters for categorical (e.g., *Gender*, *Occupation*, etc. in the political dataset; *Genre*, *Content Rating*, etc. in the movies dataset) and ordinal and numerical attributes (e.g., *Age*, *Experience*, etc. in the political dataset; *Running Time*, *IMDB Rating*, etc. in the movies dataset) can be adjusted on the left-hand side of the interface (C) using drop-down menus and range sliders. Clicking on the point in the scatterplot or on the star icon in the detail view adds the politician/movie

3 https://thestoryshack.com/tools/movie-title-generator/.

4 https://github.com/gtvalab/bias-mitigation-supplemental.

to the selected list (D). Selected data points are shown in the scatterplot with a thick red border.

Interaction Traces. In the Intervention interface, user interaction traces are shown in *real-time* in the interface with respect to *data points* and with respect to *attributes*. First, the points in the scatterplot are given a blue fill color (in-situ interaction traces) once the user has interacted with the data point, with darker shades representing a greater number of interactions (DPD metric [121]; Fig. 7.12E). The Control interface, by comparison, uses no fill color on the points (Fig. 7.12A). Second, the top right view (Fig. 7.12F) compares the user's interactions to the underlying distributions of the data for each attribute (ex-situ interaction traces). The attribute tags are colored with a darker orange background when the user's interactions deviate more from the underlying data and with a lighter orange or white background when the user's interactions more closely match the underlying distribution of data (AD metric [121]). Categorical attributes (*Gender* pictured) compare user interactions to the underlying dataset using bar charts, where gray represents the underlying distribution of data (approximately 32% women, 68% men) and a superimposed blue bar represents the distribution of the user's interactions (approximately evenly split between women and men). Numerical attributes compare user interactions to the underlying data distributions using area curves.

Real-Time versus Summative. The interaction traces pictured in Fig. 7.12(E-F) in the Intervention interface are shown in *real-time*. We also show interaction traces in a *summative* format, depicted in Fig. 7.12(G-H), after the user has made a decision (choosing 10 politicians or 10 movies). We hypothesize that both real-time and summative formats may be beneficial in different ways. In real-time, interaction traces may help users maintain awareness throughout their analysis process about the distribution of their analytic focus across the data. In a summative format, interaction traces may be easier to process and adjust from in subsequent analyses without the additional simultaneous cognitive load of the decision itself. We test variations of both in our experiment.

7.5.2 Formative In-Lab Study Results

We conducted three (3) formative in-lab studies, described in turn below. These formative studies utilized a similar task as Sect. 7.5.1 about political decision making along with earlier variants of the same control and intervention interfaces, described in Sect. 7.5.1. Analysis from these formative studies was largely qualitative and exploratory [110] in nature, informing the hypotheses and design of the confirmatory crowd-sourced experiment described in Sect. 7.5.3.

In-Lab Study 1

In the first formative study, 6 participants utilized the Control interface to choose a political committee. Our goal was to observe a baseline of user behavior and choices. Many participants intentionally balanced their political committee along several attributes (seeking "balanced representation"—P02). For example, four participants balanced by *Gender* (5 men and 5 women). The same four also balanced by *Party* (5 Republicans and 5 Democrats).

The ways that participants *biased* their committee selections were explicit but nuanced. For instance, while P05 balanced across *Gender* and *Party*, they ultimately chose a committee with all 10 members opposed to the bill, explicitly prioritizing "members (who) were very opposed to the bill." We generally observed that **participants were able to *maintain awareness* about potential biases driving their decision making**, which we hypothesized was the result of the relatively small version of the political dataset used in this study (144 data points and 5 attributes). Subsequent formative studies increased data dimensionality, from which we observed greater difficulty in maintaining conscious bookkeeping of attributes that impacted decision making.

In-Lab Study 2

In the second formative study, 12 participants each utilized the Control and Intervention interfaces to choose a political committee (24 participants in total). Our goal was to observe the effects of interaction traces on users' behavior and subsequent decisions. Exploratory analyses revealed some notable differences between participants' behavior who used the Control versus Intervention interface. In particular, for the *Age* attribute, Control participants tended to have higher Attribute Distribution (AD) bias metric values over time than Intervention participants, suggesting that **Intervention participants interacted with politicians whose ages were more proportional to the underlying dataset than Control participants** ($\mu_C = 0.857$, $\mu_I = 0.729$, $H = 3.360$, $p = 0.057$). Furthermore, participants who saw interaction traces **(Intervention) trended toward choosing more *proportional* gender composition of committees in the political task** (Fig. 7.13a); however, this trend was not replicated in the third and final formative study (Fig. 7.13b), potentially due to the introduction of confounding factors, described later.

We also observed instances where interaction traces may have led to altered behavior. For instance, after interacting with the interaction trace view, one participant's bias toward *Party* sharply decreased (as observed by the AD bias metric). One possible explanation is that the user observed bias in their interactions toward Democratic politicians in the interaction trace visualization and consequently went on to focus on Republicans to reduce the bias.

As captured by Likert ratings, participants found the *summative* metric visualization (4.5/5) more useful than *real-time* (4/5 for in-situ and 3/5 and 4/5 for categorical and numerical ex-situ representations, respectively). Participants expressed more surprise about how their interactions and selections mapped to the underlying dataset when considering the

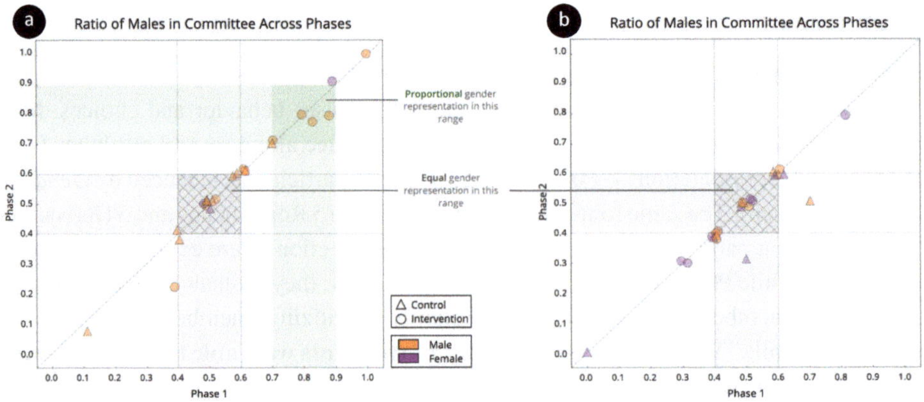

Fig. 7.13 *Gender* balance in committees chosen by 24 participants in **a** Formative Study 2 and **b** Formative Study 3. Balance is shown as the ratio of men in each participant's committee in Phase 1 (x-axis) and Phase 2 (y-axis) (shape encodes condition; color encodes participant gender)

summative view, suggesting that the view increased their awareness of bias in their analysis process (e.g., P10-I said "I'm surprised I didn't choose a doctor").

In-Lab Study 3

In the third formative study, again 24 participants utilized the Control and Intervention interfaces to choose a political committee. This study focused on qualitative analysis of *awareness*, while also addressing some shortcomings of the previous experiment (namely, the previous experiment was completed primarily by male participants, and the dataset used had only one female Republican). We observed similar, yet weaker, effects as the previous in-lab study. It could be that there is weak or no effect (which is plausible given the exploratory nature of our analyses), or it could be the result of a confounding change to the interface in this study. In particular, for the study, we permitted categorical attributes to be assigned to axes of the scatterplot. The result is that well-formed clusters appear on the scatterplot, which could itself help people more easily choose representative samples (e.g., pick a point from each cluster).

We again observed qualitative evidence of the efficacy of interaction traces toward increasing awareness of potential biases. Because the interaction trace view compares a user's interactions to the underlying distribution of the data, we hypothesized this would lead to changes in user decision making to *make the committee more proportionally representative of the underlying dataset*. For example, one participant's committee was comprised of 10 Democrats, until interacting with the interaction trace view. The participant then adjusted the committee from 10 Democrats to 4 Republicans and 6 Democrats. In fact, examination of the interaction trace view made the participant aware of a mistake in her analysis: "I forgot I had only filtered by Democrats."

Also consistent with previous findings, **participants indicated higher preference for *summative* interaction traces over *real-time* interaction traces**. Further, using a grounded theory approach to code participant utterances during think-aloud *summative* review of interaction traces, we found that **participants in the Control condition made more statements on average indicating heightened *awareness* than participants in the Intervention condition**. We hypothesize this may be due to the fact that Intervention participants already saw their interaction traces in *real-time* prior to the summative review phase.

Summary

As a result of these observations, we planned a fourth experiment to be conducted virtually with crowd-sourcing, in which we made the following adjustments:

1. we adjusted the size and dimensionality of the dataset to be sufficiently difficult such that unconscious biases may arise, and
2. we corrected for confounds resulting from interface changes between the formative studies (in the crowd-sourced experiment, we permit numerical and ordinal attributes only to be assigned to axes, while categorical attributes can be used to apply filters; implications of which are described further in the Discussion).

We also added an additional task in the domain of movies to study the generalizability of our approach. Furthermore, across all of the in-lab formative studies, the vast majority of our participants identified as Democrats. Hence, conducting an experiment via the online crowd-sourcing platform Amazon Mechanical Turk allowed us to broaden the political demographic of our users.

7.5.3 Crowd-Sourced Experiment

Procedure

This study utilized a 2 $\times$ 2 design which manipulated real-time interaction traces (present, absent) x summative interaction traces (before revision, after revision). Participants in the user study were randomly assigned to one of four conditions: CTRL, SUM, RT, or RT+SUM (Fig. 7.12I). The procedure is depicted in Fig. 7.12. After providing informed consent, participants completed a background questionnaire. Participants were shown a demonstration video of the interface using a cars dataset, then given the opportunity to practice by *choosing a shortlist of 5 cars they would be interested to test drive*.

Participants completed the first task (either politics or movies) followed by the second (movies or politics), with the order counterbalanced between subjects. For each task, participants first chose a set of 10 politicians/movies, then submitted their decision. Next, participants were either immediately given the opportunity to revise their selection (CTRL and

RT) or were shown the summative interaction trace view (SUM and RT+SUM). The summative interaction trace view shown in Fig. 7.12(G-H) depicted for each attribute of the dataset: the underlying distribution (gray), and the distribution of user interactions (blue). Then, based on any imbalances observed, participants were given the opportunity to reflect and revise their committee if desired. Lastly, those who did not see the summative interaction trace view before revision (CTRL and RT) were shown the view at the end after their decision was finalized.

Those who saw summative interaction traces *before* revision could incorporate any findings or realizations about their analysis process into subsequent revision, while those who saw summative interaction traces *after* revision could only use this information to reflect afterwards, without impacting any decisions. The study took participants 44 min on average, and they were compensated $10.

Participants

Based on a statistical power analysis from formative studies, we determined that at least 11 participants per condition would be required to detect an effect ($power = 0.8, \alpha = 0.05$). We recruited 56 participants via Amazon Mechanical Turk. We ultimately rejected 6 submissions due to a combination of failed attention checks, missing data, speeding through the study, and poor open-ended responses, leaving us with data from 50 participants who were randomly assigned to one of four conditions (13 CTRL, 14 SUM, 11 RT, 12 RT+SUM). Workers were restricted to only those located in the U.S. state of Georgia with 5,000+ approved HITs over their lifetime and a $\geq 97\%$ approval rating. By gender, participants identified as female (28), male (21), and agender (1). Participants were 24–69 years old ($\mu = 40$, 1 preferred not to say) and self-reported an average visualization literacy of $\mu = 3.08, \sigma = 0.9$ on a 5-point Likert scale. They had a wide range of educational backgrounds (6 high school, 12 some college, 7 associate's degree, 20 bachelor's degree, 3 master's degree, and 2 post-graduate degree); and a variety of fields of work, including e.g., art, bookkeeping, computer science, criminal justice, marketing, microbiology, office administration, political science, sales, social work, among others. We refer to participants from each condition as $\{P_{CTRL}1$ - $P_{CTRL}13\}$, $\{P_{SUM}1$ - $P_{SUM}14\}$, $\{P_{RT}1$ - $P_{RT}11\}$, and $\{P_{RT+SUM}1$ - $P_{RT+SUM}12\}$.

Politics Background. Most participants had voted in US Presidential (49), state (43), or local (35) elections. Participants identified as Democratic (35) and Republican (15), rating themselves on the political spectrum as conservative (2), moderate conservative (9), moderate (8), moderate liberal (14), and liberal (17).

Movies Background. Participants rated varying importance of movies in their lives (2 no importance, 15 little importance, 20 moderate importance, 12 large importance, 1 most importance). They watched movies daily (4), weekly (30), or monthly (16) and reported a diverse range of preferred genres.

Hypotheses

Based on findings from formative in-lab studies, our hypotheses for this experiment are as follows. We organize our hypotheses according to those regarding **B**ehavior, **D**ecisions, **A**wareness, and **U**sability.

B1 *Real-time* interaction traces will have an effect on users' analysis *process*.
B2 *Summative* interaction traces seen *before* revision will lead users to make more *revisions*.
D1 SUM, RT, and RT+SUM participants will make selections more proportional to the underlying data than CTRL participants.
A1 CTRL participants will exhibit greater *surprise* upon seeing *summative* interaction traces (Fig. 7.12(G-H)) than participants in SUM, RT, and RT+SUM.
A2 The attributes for which participants indicate *surprise* about their interaction traces will correlate to lower *focus*.
A3 The attributes for which participants indicate *surprise* or *focus* about their interaction traces will correlate to AD metric values.
U1 Participants in the RT and RT+SUM conditions will *not* consistently use the real-time interaction trace view (Fig. 7.12F).
U2 Participants will find the summative interaction trace visualization (Fig. 7.12(G-H)) more useful than real-time interaction trace visualizations (Fig. 7.12(E-F)).

Based on guidance for statistical communication [37], our analyses relied primarily on parameter estimation for all quantitative measures, using empirical bootstrapping with 1000 resamples to estimate the 95% confidence intervals around all sample means. We prioritize reporting results for attributes that users indicated as high focus (e.g., the top three are *Ban Abortion After 6 Weeks*, *Party*, and *Gender* for the political task; *IMDB Rating*, *Genre*, and *Rotten Tomatoes Rating* for the movies task). Complete analyses are available in supplementary materials.

Behavior

We hypothesized that the presence of *real-time* interaction traces would impact user behavior as measured by (1) interaction counts, (2) bias metric values [121], and (3) revisions during Phase 2.

Interactions. Figure 7.14 (left, center) illustrates total number and type of interactions performed by users in each condition for both the politics and movie tasks. We find some notable distinctions as observed by lesser overlap in confidence intervals. Namely, CTRL participants performed fewer hover interactions and fewer total interactions, demonstrating less interactive behavior than those who saw interaction traces. However, other specific interaction types showed less obvious trends. This result provides **some support for hypothesis B1**.

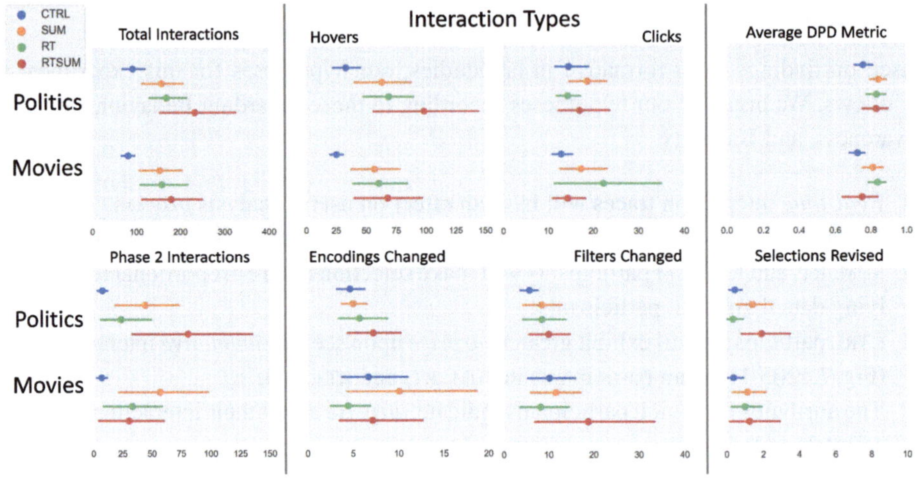

Fig. 7.14 Modeling user behavior using point and interval estimation of the mean value of interactions of different types performed, revisions made, and the DPD bias metric

Bias Metrics. The AD bias metric values provide one way of quantifying how a user's interactive behavior aligns with the distributions of the underlying data per attribute [121]. Lower metric values indicate interaction distributions that are more similar to the distribution of a given attribute, while higher metric values indicate dissimilarity. Figure 7.15 (left) shows the average AD bias metric values for the top three attributes that participants focused on in each task. We observe that CTRL participants exhibited less bias toward some attributes (e.g., *Gender* and *Party* in the political task and *Genre* in the movies task) while other attributes display less clear trends.

The DPD bias metric similarly quantifies how evenly a user's interactions are divided among individual data points. Figure 7.14 (top, right) shows a slight trend toward lower DPD metric values for CTRL participants. These results provide **mixed support for hypothesis B1**.

Revisions. All participants had the opportunity to revise their initial selections. We hypothesized that those participants who saw *summative* interaction traces *before* revision would make more revisions (i.e., number of edits to their initial selection after Phase 1). Figure 7.14 (bottom right) shows that SUM and RT+SUM participants who saw summative interaction traces before revision tended to make more revisions compared to CTRL and RT participants during the political task, but not during the movies task. Similarly, the same groups tended to perform more interactions in general in the phase 2 revision (Fig. 7.14 (bottom left)), demonstrating that summative interaction traces tended to correspond to more exploration and decision changes during revision. This result **confirms hypothesis B2**.

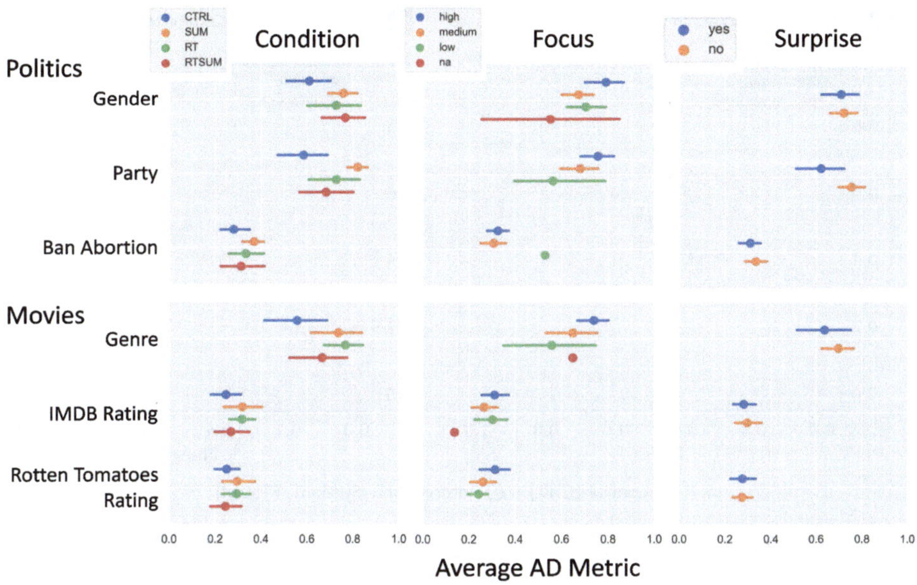

Fig. 7.15 The effect of condition, focus, and surprise on estimating the average AD metric value for select attributes in each task

Decisions

We hypothesized that participants in SUM, RT, and RT+SUM (i.e., those who were influenced by interaction traces in some format, real-time or summative, before finalizing their decisions) would ultimately make choices that were more proportional to the underlying data compared to participants who did not (CTRL). We quantify this effect for binary attributes in the datasets (e.g., *Party* and *Gender* in the politics dataset), by considering the ratio of values *chosen* with respect to the ratio that appears in the *dataset*. In the movies task, no attributes are binary; hence this analysis only applies to *Party* and *Gender* in the politics task.

For *Party*, the dataset contains 59% Republicans and 41% Democrats. For *Gender*, the dataset contains 32% females and 68% males. As shown in Fig. 7.17, *all conditions* chose committees that were relatively dissimilar from the underlying distributions of *Gender* and *Party* in the dataset (annotated with a vertical dashed line). We discuss this result further in the next chapter. In fact, for participants who saw any intervention (RT, SUM, RT+SUM), they trended toward choosing *more dissimilar* distributions of *Party* compared to CTRL participants. However, there are no clear distinctions in the ratios of *Gender* or *Party* between conditions. Hence, we find **no support for hypothesis D1**.

While we observe no clear effects on committee composition based on intervention conditions, we do see some distinctions in the way participants chose their committee based on their own political party affiliation. For instance, Fig. 7.16a shows the ratio of Democrats that

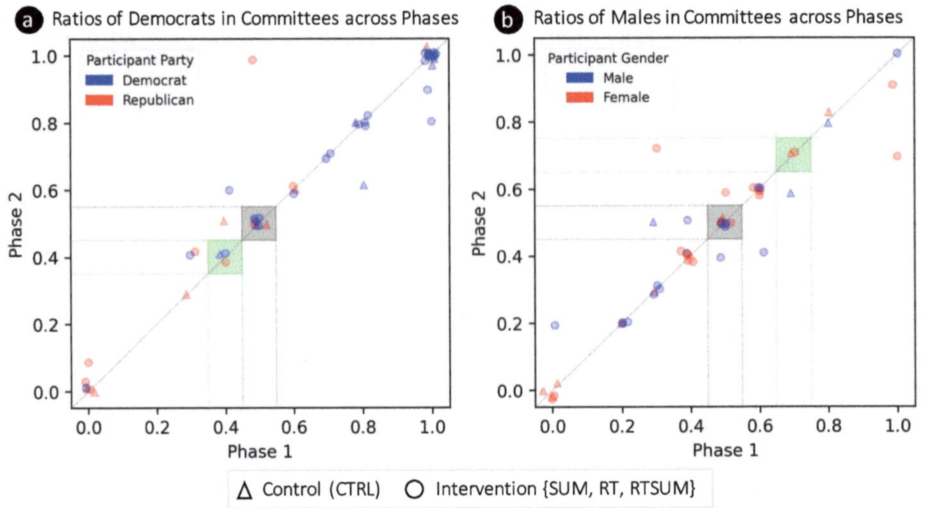

Fig. 7.16 Ratio of *Party* **a** and *Gender* **b** composition of committees in Phase 1 and Phase 2 of the political task, colored by the participant's corresponding party affiliation and gender

participants chose in Phase 1 (x-axis) and in Phase 2 (y-axis) committees. Points that fall on the diagonal represent participants who did not change the composition of their committee by *Party* during the revision. Points are colored by the political party affiliation of the participant. There is clear delineation, where participants who most identified with Democrats (blue) appear in the top right (choosing Democrat-dominant committees) and participants who most identified with Republicans (red) appear in the bottom left (choosing Republican-dominant committees). Three notable examples emerge e.g., a blue circle in the bottom left, a red circle in the top middle, and a red triangle in the top right. These participants chose final committees that were entirely composed of politicians from the opposite *Party* than their own affiliation. Upon further inspection, two participants identified as moderate or neutral on the spectrum of liberal → conservative, and one expressed divergent beliefs from their affiliated party. For *Gender* (Fig. 7.16b), we do not observe such a clear distinction. Participants' committees reveal some clusters (e.g., in the bottom left, many female participants chose committees of all female politicians; just above that cluster along the diagonal, many men chose committees with 70–80% female politicians; etc.). However, there is less strict division in the overall trend ($\mu_{Female} = 0.45$, $\mu_{Male} = 0.44$).

Awareness

While formative studies indicated variable impacts of interaction traces on behavior and decisions, they had a more promising qualitative effect on people's *awareness* of potential unconscious biases that drive data analysis and decision making. In particular, we assess

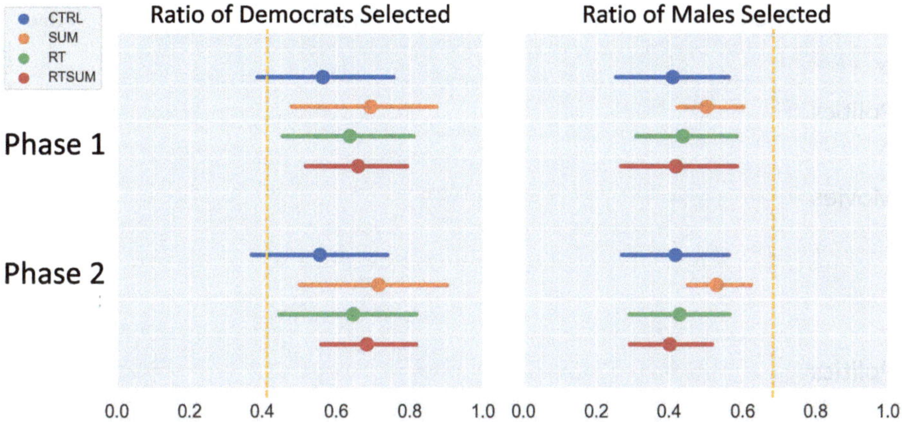

Fig. 7.17 Mean estimates for the ratio of Democrats (left) and Males (right) in participants' chosen committees, selected during each phase for each condition. Vertical dashed lines indicate ratio of each candidate type present in the underlying dataset

awareness by asking two questions for each attribute of the data, at the time that the user viewed the summative interaction traces (*before revision* for SUM and RT+SUM, and *after final selections* for CTRL and RT).

1. Are you **surprised** by your interactions with this attribute? {*yes*, *no*}
2. How much **focus** did you give this attribute during your task? {*high*, *medium*, *low*, *NA*}

Figure 7.18 compares the average number of times a particular combination of focus and surprise was recorded between all conditions and tasks. We hypothesized that CTRL participants would express surprise more often upon seeing the *summative* interaction traces than participants in the other conditions. Participants in all other conditions had some form of signal from interaction traces (real-time or summative) *before* their final decisions, whereas CTRL participants only saw the summative view after their final selections were locked in. In the political task, we observe that CTRL participants expressed surprise for high-focus attributes more often than the other conditions, and conversely reported the lowest numbers of no surprise for high-focus attributes (Fig. 7.18 left). This provides some support for the idea that seeing summaries, either in real-time (RT), before revisions (SUM) or both (RT+SUM), reduces the surprise reported when participants expressed high focus in more divisive analysis scenarios. The effect was less pronounced in the movies task. In general, as focus decreased on any given attribute, the number of surprised responses decreased uniformly as well across conditions and tasks. This pattern further supports the idea that summaries can affect whether participants are aware of their analysis strategies at the end of a task. This result **confirms hypothesis A1**.

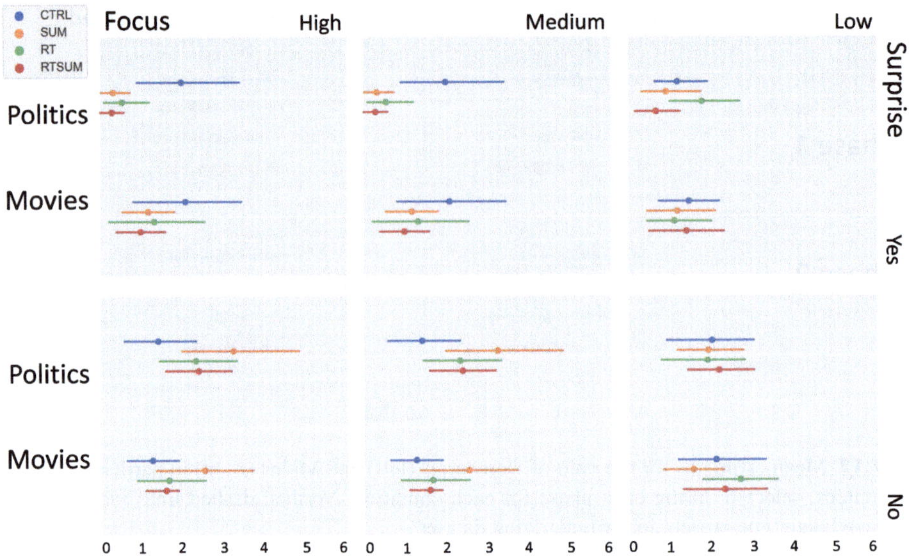

Fig. 7.18 Estimating the mean number of times a combination of focus and surprise was recorded for each task and condition

We also hypothesized that expressing lower focus on an attribute (e.g., *Ban Abortion After 6 Weeks*, *Party*, and *Gender* for the political task; *IMDB Rating*, *Genre*, and *Rotten Tomatoes Rating* for the movies task) would be correlated to more instances of surprise. That is, we believed attributes that were unattended may have a surprising distribution of user interactions, since the user did not focus on them. Comparing each column of Fig. 7.18 top versus bottom demonstrates that when CTRL participants expressed surprise (top, blue), fewer people expressed that level of focus compared to when participants were not surprised (bottom, blue). For other conditions, there does not appear to be any substantial difference in surprise (top) versus no surprise (bottom) w.r.t. focus. That is, surprise and focus do not appear to be correlated. All means and intervals are roughly between 1 and 3. This result **disconfirms hypothesis A2**.

Lastly, we hypothesized that there would be a correlation between attributes that participants' *focused* on (Fig. 7.15 center) or were *surprised* by (Fig. 7.15 right) and the average AD bias metric values. Some attributes (e.g., *Party* in the political task, *Rotten Tomatoes Rating* in the movies task) roughly corresponded to greater focus related to higher AD bias metric values. On the other hand, some attributes for which participants expressed surprise about their interactions (e.g., *Party* in the political task) corresponded to lower AD bias metric values However, these trends were not true for all attributes, hence **support for hypothesis A3 is inconclusive**.

Usability

Based on formative studies, we formulated two general hypotheses about the usability of interaction trace interventions. First, in formative studies we observed relatively little use of the *real-time* interaction traces in the form of interactions with the Distribution panel (Fig. 7.12F), which we believe to be due to high cognitive load during the task itself (i.e., participants were unable to attend to an additional view in the system while trying to explore the data). Indeed, participants interacted minimally with the Distribution panel in both the politics ($\mu_{RT} = 2.09$, 95% CI [0.64, 3.82], $\mu_{RT+SUM} = 1.33$, 95% CI [0.33, 2.42]) and movies tasks ($\mu_{RT} = 3.18$, 95% CI [0.45, 6.28], $\mu_{RT+SUM} = 2.08$, 95% CI [0.17, 4.67]). This result **confirms hypothesis U1**.

For similar reasons (high cognitive load), we hypothesized that participants would prefer the *summative* interaction traces over the *real-time* interaction traces. This result was somewhat variable. CTRL participants saw only summative interaction traces (Fig. 7.12G-H, after revision) and rated their utility on a Likert scale a median 4/5 across both politics and movies tasks. SUM participants who saw only the summative interaction traces (before revision) likewise gave a median 4/5 Likert rating for movies and 3.5/5 for politics. RT participants rated in-situ interaction traces (Fig. 7.12E), ex-situ interaction traces (Fig. 7.12F), and summative interaction traces (after revision) all the same with a median Likert rating of 4/5 across both politics and movies tasks. These participants all viewed *summative* interaction traces very positively. Only RT+SUM participants differed, rating in-situ interaction traces 4/5 (politics and movies), ex-situ interaction traces 2/5 (politics) and 3/5 (movies), and summative (before revision) 3/5 for both politics and movies. The participants in the only condition that could compare both real-time and summative interaction traces surprisingly preferred *real-time* interaction traces. Overall, RT+SUM participants rated all forms of interaction trace lower than participants in the other conditions, perhaps due to this condition having the highest cognitive load of all the conditions, showing both real-time and summative interaction traces. This result **disconfirms hypothesis U2**.

Qualitative Feedback

The survey after each task included open-ended questions about the participant's decision criteria and, in the political scenario, their desired outcome of the committee. Below we discuss themes that emerged.

Politics. Some focused on choosing diverse committees (e.g., $P_{CTRL}2$ said "[they] tried to choose politicians with a wide range of views, to represent most people's views and not necessarily one side"). Many expressed firm goals about choosing politicians differently than the underlying data distribution (e.g., $P_{CTRL}1$ expressed they wanted their committee to be mostly women because "[they] truly feel abortion is an issue only women can really comment on"), while also simultaneously balancing other attributes (e.g., "[they] also tried to get a mix of Republicans and Democrats with many years experience in politics to get a fair showing to both sides"). Others sought committee members based on "who [they] thought might share [their] values" ($P_{CTRL}10$), instances of which were observed from both sides of the political spectrum (e.g., $P_{SUM}8$ chose "[Republican] politicians who were in favor of the abortion ban", while $P_{SUM}13$ intentionally "chose all Democrats").

Ultimately, people hoped their chosen committee would lead to outcomes such as "uphold the ban on abortion after 6 weeks, except in extreme cases where there is a life-threatening decision that needs to be made" ($P_{RT}6$), that "the abortion ban would become or stay law" ($P_{SUM}12$), or "assess the public opinion in an unbiased way" ($P_{RT+SUM}2$). Many of the biases that emerged as a result of these goals were very conscious. For instance, $P_{RT}3$ expressed, "I was really biased, to be honest. I wanted people who were in favor of abortion, and most, if not all, Republicans did not fit the bill." In some cases, participants adjusted their strategy for Phase 2 (Revision); e.g., $P_{RT+SUM}6$ said, "my revised committee was solely focused on trying to get vast ideological perspective without necessarily focusing on men or women. I did more of a 'blind' choosing without focusing so much on gender the second time around ... I only wanted different viewpoints without focusing too much on the extreme of either side."

Movies. Participants expressed diverse criteria for selecting movies. Some focused on making choices that "were a good representation of the given dataset as a whole" by "find[ing] and select[ing] films that were spread out across the graph" ($P_{CTRL}7$). Several other participants expressed a focus on finding variety for one factor, e.g., highly rated ($P_{CTRL}6$) or diverse genres ($P_{CTRL}5$). Others had criteria that were too abstractly expressed to capture in the data (e.g., "selecting movies that [they] feel would be interesting to watch" -$P_{CTRL}11$). Some participants expressed a focus on the dummy *Title* attribute ($P_{RT}7$).

Some participants expressed how interaction traces influenced their behavior. For instance, in Phase 2 (Revision), $P_{SUM}6$ expressed that "the 2nd time through, [they were] careful to check the different areas such as genre" to "[assess] which [they] thought were representative in each case." Similarly, $P_{SUM}13$ was "glad [they] got to go back and revise because [they] missed the creative type selection which [they] corrected to contemporary fiction." Others relied on different features to facilitate their analysis, e.g., filters ($P_{RT}3$).

7.6 Research: RutaBAGA—A Case Study on Metacognitive Awareness in University Graduate Admissions

We introduced interaction traces through the Lumos system in Sect. 7.4 and showed how the technique can lead to increased awareness of biases in Sect. 7.5. Next, we show how the same technique can lead to mitigation of implicit biases in the context of university admissions through a controlled study and case study.

University admissions involves complex decision-making processes, often characterized by individual reviewers reading application packets and rating them based on factors such as the applicant's academic performance, non-academic accomplishments, and personal qualities such as communication skills. Reviewers' individual evaluations are often then collated and discussed among a committee to inform final admissions decisions. Given the complexity and subjectivity of this process, **unconscious biases** (biases beneath the level of an individual's conscious attention) might influence reviewers' decisions.

This work focuses on a specific type of unconscious bias—implicit biases, which can include gender and racial bias, that are ingrained as a result of cultural norms or past experiences [53, 54]. Unlike conscious biases such as overt sexism or racism which involve deliberate prejudice against people of a certain group, implicit biases are not intentional and may not always align with one's explicit beliefs [103]. Nevertheless, they often materialize in harmful ways such as snap judgments based on an individual's skin color or gender. Importantly, implicit biases are not conscious, they often persist despite an individual's best efforts to counteract it. Our goal in this work is therefore to investigate (i) to what extent implicit biases correspond to observable behaviors in the review process, (ii) how we can promote real-time reflective review processes, and (iii) to what extent heightened awareness leads to reviewers' adjustment of associated behaviors and decisions.

We designed an interactive system, **RUTABAGA**, to support a bias aware graduate admissions process. We introduced the intervention in the context of graduate admissions and later generalized the concept to undergraduate admissions as well but retained the system name. In this chapter, we define **bias** to encompass inequities in (i) the review process (via disparities in time spent reviewing applicants of different racial and gender groups) and (ii) decisions (via disparities in perceived competitiveness of applicants of different racial and gender groups). We aim to achieve **awareness** by surfacing these disparities to reviewers through visualizations to encourage real-time **self-reflection** on the review process.

This work was originally published in the Journal of Visualization and Interaction (JoVI) in 2025 [33]. We emphasize the following contributions:

1. the system, RUTABAGA, designed in collaboration with two Ph.D. program admissions committee chairs to promote reflection on review processes and increase reviewer awareness of undesired behaviors,
2. empirical results demonstrating (i) the relationship between implicit biases and observable review behaviors and (ii) the impact of RUTABAGA on bias awareness, and
3. results of a case study with the Computer Science department at a private university that demonstrate how RUTABAGA can facilitate bias-aware decision making in a real-world system deployment.

We discuss the implications of our findings, and in particular, how they relate to the 2023 Supreme Court decision overturning affirmative action that ruled universities may no longer make admissions decisions on the basis of race [113]. We conclude with a call-to-action—that research to address biases in admissions processes has never been more urgent.

7.6.1 Formative Design

The design and development of RUTABAGA followed a user-centered approach [2] that involved close collaboration with two Ph.D. admissions committee chairs in the Computer Science Department of a private university.

7.6.1.1 Methodology

We first conducted **semi-structured interviews** with two admissions committee chairs, C1 and C2, in the Computer Science department to understand their Ph.D. admissions process in detail. The 30-minute interview covered topics including the application format, data access, decision-making criteria, and collaborative mechanisms. The interviews illuminated **characteristics of the existing review process** (Sect. 7.6.1.2) and the program's need to conduct meta-analysis on the admissions review process.

Based on our understanding of the program needs, we next sketched several solutions and built a **preliminary prototype** to ground the discussion in a second interview. More details of the preliminary interface are in Supplemental Materials. In the second interview, we provided a **demonstration of the interface design** to the chairs, followed by a semi-structured interview to gauge initial reactions to the design. The session lasted approximately 45 min. An additional survey followed afterward, including questions about the usefulness of system components, willingness to use such a system in the future, and an opportunity to suggest alternatives in free-form text. The main feedback we received included the chairs' desires to (1) visually incorporate interaction *time* on an application to allow assessment of time spent across applicants (the preliminary system included visual encoding of discrete interaction *count*) and (2) make the interface simpler to use by separating the review phase and self-reflection phase into separate tabs which were initially adjacent views in a singular screen in the prototype.

Additional interviews were conducted with one committee chair (C1) after two additional iterations of design to collect **ongoing feedback on subsequent iterations** of the system. Each session lasted about 30 min. The system was updated according to feedback including (i) adding visual encodings for applicants' gender and race and (ii) visualizing reviewers' time spent on each *component* of an application, as described in Sect. 7.6.2. From the interviews and questionnaire responses, we summarized the needs for analyzing the admissions review process as **process awareness needs** (Sect. 7.6.1.3) and refined our **design goals** (Sect. 7.6.1.4) for the final system design.

Following the initial deployment of rutaBAGA, the system was further refined based on observations of its real-world usage. These ongoing refinements are discussed in Sect. 7.6.7.

7.6.1.2 Prior Review Process

The interviews illuminated the prior review process which can be summarised as follows. Applicant portfolios, in the form of single PDF file (one per applicant) were distributed to a minimum of two reviewers assigned by the committee chairs based on applicant research interests. Each reviewer rated the assigned applicants according to pre-defined criteria such as Research Preparedness and Communication Proficiency on a 0–5 scale and shared optional free-form feedback in a shared spreadsheet. After the initial review, applicants who were "above the bar" were interviewed via Zoom by at least one faculty member. The committee then met to discuss and make admissions decisions, referencing a sorted version of the spreadsheet of reviewer scores to anchor discussions.

7.6.1.3 Process Awareness Needs

Based on the interviews, we identified reviewers' needs for investigating (i) internal consistency in *time spent* across applicants and application components, and (ii) internal consistency in *ratings and decisions* across applicants, stratified by attributes such as race and gender. Inconsistency in time spent (i), for instance, could be not spending enough time on a certain application, neglecting a certain application component (e.g., not reading recommendation letters), or systematically spending more/less time on a certain group of applications across gender or race. While time spent alone is a noisy metric influenced by many factors, as suggested by the admission chairs, it would be a useful metric that can nonetheless provide reviewers with some point of reference to spark reflection.

The system should also encourage reviewers to reflect on the decision outcomes and check internal consistency in their ratings (ii). Inconsistency in ratings could be that applications with the same ratings in evaluation criteria received different overall ratings (e.g., competitive v. not competitive), or that a certain group of applications was systematically rated as more/less competitive across gender or race, etc.

7.6.1.4 Design Goals

Based on our formative design activities, we derived the following **design goals (DG)** to support the program's desired framework for admissions review needs for increasing process awareness.

DG1. Facilitate independent review of applications. The system should enable reviewers to *independently* rate applications across pre-defined dimensions.

DG2. Support assessment of independent review behavior. The system should enable individual reviewers to analyze their decision-making processes and outcomes to increase awareness of undesired behaviors, such as inconsistency in time spent across applications.

DG3. Minimize the barrier to entry. The system should be visually simple and intuitive to increase adoption of the system over the status quo method for reviewing, and to ensure the system is usable by faculty members beyond the visualization domain.

The system was also designed with goals in mind to support collective *group* awareness and decision making; however, we defer discussion of these features to supplemental materials as they were not the focal point of our subsequent experiments.

7.6.2 System

Based on our design goals, we developed a system, RUTABAGA (Fig. 7.19), mainly consisting of two separate tabbed pages, including a Rating Page where reviewers read and rate applications independently (DG1); and a Summary Page where reviewers can see a summary of their process as shown from their interactions with applicants (DG2).

7.6.2.1 Rating Page

The Rating Page (Fig. 7.19, top) is designed to support seamless completion of existing tasks involved in independent review of applications (DG1). It consists of the following components.

(A) Document Viewer shows PDF documents including personal statement, resume, letters of recommendation, transcript, and so on, organized into tabs. **(B) Profile View** shows tabular attributes of the applicant such as GPA, degree, major, etc. A set of default attributes are shown initially, and users can select/deselect attributes to be shown from a drop-down list if they would like to make their decision process blind to some sensitive attributes. **(C) Comments View** allows reviewers to leave comments about the respective application. **(D) Ratings View** allows reviewers to (i) rate the applicant on a set of factors (which can be pre-defined by the committee based on their review criteria) on a 0–5 scale and (ii) rate the overall competitiveness of the applicant on a 1-4 scale (1 = Not Competitive, 4 = Very Highly Competitive). Composing the application documents, comments, and ratings in a single interface makes it simpler to evaluate applications and adopt the system (DG3).

Interaction Logs record reviewers' time-stamped interactions including clicking and scrolling on each applicant and each component of the application (DG2). Page visibility changes (i.e., navigating to a different application or browser tab) are recorded in order to facilitate filtering out time periods that users are not focused on the interface. Furthermore, the system applies thresholds to filter out outlier time periods (that are too short or too long) in order to reduce noise in derived time spent. Inspired by prior work in analytic provenance [88], which demonstrates that user interaction is a powerful cue to learn about users, we used the interaction logs to derive visualizations of reviewers' review process to help them maintain self-awareness of their decision-making process.

Fig. 7.19 RUTABAGA supports admissions review with two pages. 1. Rating Page: (A) Documents Viewer shows different application documents, (B) Profile View shows a set of attributes and a drop-down list for selecting/deselecting visible attributes, (C) Comments View and (D) Ratings View. 2. Summary Page (case study, middle) maintains the right-hand-side of the interface for profile, comments, and ratings of individual applicants, and replaces Document Viewer with (E) Filters, (F) Interactive Scatterplot, and (G) Time Spent Distribution Chart. The Summary Page was modified for the controlled study (bottom) which shows strip plots representing users' Time Spent by Gender (H, top) and Race (H, bottom), and Competitiveness Rating by Gender (I, top) and Race (I, bottom) as well as the full applicant list (J). Note that all the application material displayed here is synthetic data for demonstration purposes

7.6.2.2 Summary Page

While the Rating Page serves as an interface for completing existing reviewing tasks, the Summary Page (Fig. 7.19, middle) aims to enhance awareness of individual reviewers' processes (DG2). Reviewers can access this page at any time during the admissions review cycle. It maintains the **Profile, Comments,** and **Ratings** Views (Fig. 7.19 B, C, and D, respectively) from the Rating Page, but replaces the **Document** (A) panel with a data visualization panel (A.1), as shown in the middle of Fig. 7.19.

(E) Filters provide controls for filtering data by numerical or categorical attributes. Beyond the Profile attributes (gender, test scores, etc.), users can also filter by their assigned ratings (for teaching preparedness, communication, etc.) and overall recommendation of applicants (e.g., to view only candidates they rated as Competitive).

(F) Interactive Scatterplot visualizes reviewed applications, with the ability to assign x- and y-axes from a drop-down list to represent variables like GRE score, GPA, reviewer's ratings, and overall recommendation. *Hovering* on a point (applicant) in the scatterplot populates the Profile, Comment, and Ratings Views, and the Time Spent Distribution (described below) with the applicant's data, and *clicking* a point returns to the applicant's portfolio in the Rating Page. Additionally, the scatterplot allows for additional encodings of size (time spent) and color (overall recommendation of an applicant, applicant gender, or applicant race).

The scatterplot is designed based on the need for providing visualizations that allow reviewers to observe patterns and outliers in their time spent and ratings (DG2). While there are many visualization techniques for representing multi-dimension data such as parallel coordinates and scatterplot matrix, we chose a scatterplot because of its effectiveness in identifying patterns/outliers, visual simplicity, and familiarity to general audiences [76] (DG3).

(G) Time Spent Distribution shows a grouped bar chart depicting (i) the reviewer's average time spent on different documents and (ii) the reviewer's distribution of time spent on application components for the hovered applicant in the scatterplot. This view is designed to help users gain insights about the time they spent across different application components (DG2) and identify outliers at the individual applicant level such as little or no review of a certain file for an applicant.

7.6.3 Controlled Study Methodology

We conducted a pre-registered[5] crowdsourced study to assess (i) the underlying assumption of RUTABAGA that implicit biases (as measured by implicit association test scores [55]) correlate to observable differences in application review behaviors and decisions and (ii) the effectiveness of the RUTABAGA in enhancing awareness of undesired behaviors. Participants

[5] https://aspredicted.org/blind.php?x=N51_KWZ.

conducted a simulated review of 12 undergraduate university applications using a variation of RUTABAGA described below (Sect. 7.6.3.3).

7.6.3.1 Participants

We initially recruited 75 participants through the Prolific crowdsourcing platform based on power analysis (with $\beta = 0.05$, $\alpha = 0.8$) from a pilot study. Seven participants who had invalid gender or race IAT scores (due to rapid responses) were excluded, and we recruited additional participants to maintain our target sample size of 75 (31 identified as female, 40 as male, and 4 as non-binary). Eligibility criteria included 18+ years old, hold a Bachelor's degree or higher, and fluent in English. Twelve participants indicated prior experience in university admissions. Participants spent 77 min on average ($min = 25$, $max = 162$) and were compensated $15.

7.6.3.2 Dataset

The dataset used in this experiment was comprised of 12 artificially generated university application packets. While we would ideally explore a diverse range of racial and gender identities, we simultaneously were constrained to limit the average study duration to 1 h to minimize participant fatigue. Therefore, we opted to study three levels for gender (non-binary, female, male); two levels for race (Black, White) [15]; and two levels for competitiveness (high, moderate), resulting in 12 unique application portfolios ($3 \times 2 \times 2$) as summarized in Table 7.1. Each application packet consisted of a personal statement, resume, transcript, two recommendation letters, and basic form-field information including GPA, SAT score, class rank, gender, and race. We used ChatGPT to generate the application packets using a sequence of prompts shared in Supplemental Materials.

Table 7.1 A summary of the 12 artificially generated applications

#	Name	Gender	Race	Competitiveness	SAT Score	GPA	Class Rank
1	Aaliyah Washington	Female	Black	High	1500	3.9	5/100
2	Keisha Brooks	Female	Black	Moderate	1290	3.4	90/150
3	Emily Johnson	Female	White	High	1550	3.8	14/200
4	Sarah O'Donnell	Female	White	Moderate	1280	3.4	163/250
5	Jamal Richardson	Male	Black	High	1540	3.9	12/200
6	Terrence Brown	Male	Black	Moderate	1260	3.2	68/100
7	Lucas Miller	Male	White	High	1490	3.7	14/150
8	Daniel Smith	Male	White	Moderate	1300	3.5	180/250
9	Taylor Patterson	Non-binary	Black	High	1520	3.8	16/200
10	Jordan Thomas	Non-binary	Black	Moderate	1300	3.5	183/250
11	Alex White	Non-binary	White	High	1530	4.0	1/100
12	Casey Sullivan	Non-binary	White	Moderate	1270	3.3	105/150

7.6.3.3 Interface

We created a variation of RUTABAGA for the controlled study which included the Rating page (as described in Sect. 7.6.2.1) and a modified version of the Summary page (as described in Sect. 7.6.2.2). The Summary page (Fig. 7.19, bottom) was updated primarily to adjust features that do not scale down well for the study where only 12 applicants were rated. Namely, we removed the filters (E) and replaced the scatterplot (F) and bar chart (G) with strip plots (H, I) that showed the distribution of users' time spent and overall ratings across different gender/race groups. Each circle on the strip plot represents an applicant and the vertical line indicates the mean value. We added an applicant list (J) to allow easy navigation between applicants with brushing and linking between views. The Ratings panel (D) was also updated to a slider ranging from 1–100 that allowed participants to give an overall competitiveness score for an applicant for a more granular analysis.

7.6.3.4 Procedure

This study utilized a within-subjects design where participants first reviewed the applications without the intervention (Control) and were then directed to the Summary Page (Intervention) after rating all the applications. This design allowed us to account for individual differences in review behaviors and decisions.

Participants first completed a step-through tutorial that introduced the system's layout for displaying application portfolios and operations for entering ratings. After completing the walk-through training, participants reviewed 12 applications (order randomized) and rated them on a scale of 1 (not competitive) - 100 (very competitive). After participants rated all the applications, they completed a second step-through tutorial that introduced the features in the modified Summary Page (Fig. 7.19, bottom). Participants were able to interact with the page and revisit profiles to finalize their ratings. The post-study survey included background information such as gender, education level, prior experience in university admissions, and Likert-scale questions about the summary page. Finally, participants completed the Gender-Science and Race implicit association tests (IAT) adapted from Project Implicit [92] (order randomized).

7.6.3.5 Hypotheses

We organize our hypotheses into three groups:

H1: Participants' review behavior (time spent) and decisions (competitiveness ratings) will differ based on applicants' race and gender.

- H1.1. Female and non-binary applicants will receive less *review time* than male applicants.
- H1.2. Female and non-binary applicants will be rated as *less competitive* than their equally-qualified male counterparts.
- H1.3. Black applicants will receive less *review time* than White applicants.

- H1.4. Black applicants will be rated as *less competitive* than their equally-qualified White counterparts.

H2: Participants' review behavior (time spent) and decisions (competitiveness ratings) will correlate with their implicit biases.

- H2.1. Participants with higher gender IAT scores will spend *less time* on female and non-binary applicants than male applicants.
- H2.2 Participants with higher gender IAT scores will give *lower ratings* to female and non-binary applicants than male applicants.
- H2.3 Participants with higher race IAT scores will spend *less time* on Black applicants than White applicants.
- H2.4 Participants with higher race IAT scores will give *lower ratings* to Black applicants than White applicants.

H3: Visualizations of the review process will affect reviewers' behavior and decisions.

7.6.4 Controlled Study Results

In the forthcoming analyses, we normalized participants' time spent and competitiveness ratings to fall between 0 and 1 to control for individual variations. For the analyses of H1 and H2, we used participants' time spent and competitiveness rating data before they interacted with the summary page (Control), while the analysis of H3 considered interactions both before and after the summary page (Intervention).

7.6.4.1 H1: Effects of Applicant Gender and Race

We stratified our analysis of time spent and competitiveness ratings by applicant race and gender. Figure 7.20 summarizes the normalized time spent reviewing applicants (left) and normalized competitiveness ratings (right) for applicants by gender (top) and race (bottom). Each depicts a box plot representing the 75 participants' time spent or ratings for each applicant that falls into the respective gender or race group. We used a non-parametric test (Mann-Whitney U test) instead of the pre-registered t-test for the analysis since the data do not follow a normal distribution as assessed by the Shapiro-Wilk's test ($p < 0.05$).

We found **partial support for H1.1**. Female applicants ($Mdn = 0.209$) received less review time than male applicants ($Mdn = 0.253$) ($p = 0.004$). On the other hand, non-binary applicants ($Mdn = 0.299$) received more review time than male applicants ($Mdn = 0.253$); however, the difference is not statistically significant ($p = 0.156$). Additionally, we found **no support for H1.2:** female ($Mdn = 0.566$) and non-binary applicants ($Mdn = 0.555$) were rated less competitive than male applicants ($Mdn = 0.633$); how-

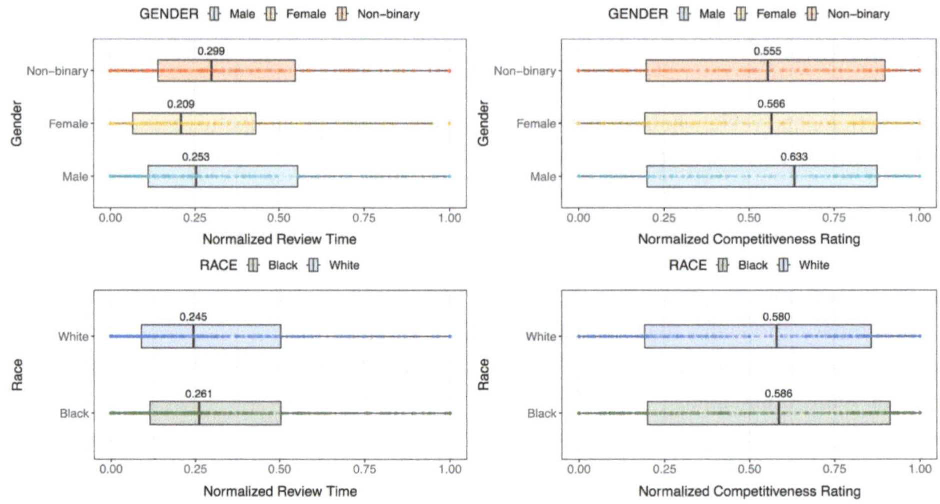

Fig. 7.20 Boxplots with individual points of normalized review time (left) and normalized competitiveness ratings (right) by applicant gender (top) and race (bottom). The values are normalized per participant, where 0 represents the minimum review time/rating, and 1 represents the maximum review time/rating to facilitate a comparison that accounts for individual differences

ever, the differences lack statistical significance ($p = 0.656$ and 0.920, respectively). We also found that participants spent more time reviewing Black applicants ($Mdn = 0.261$) than White applicants ($Mdn = 0.245$) and gave them higher average ratings ($Mdn = 0.586$) than White applicants ($Mdn = 0.580$) . These results contradict the directionality of our hypothesis H1.3 and H1.4, but also lack statistical significance ($p = 0.140$ and 0.286, respectively).

We further validated that findings for H1 were not influenced by order effects. Details of the analysis are included in Supplemental Materials.

7.6.4.2 H2: Implicit Biases

To investigate the extent to which implicit biases correspond to deviations in time spent and competitiveness ratings for applicants of different races and genders, we conducted linear regression analyses on the race and gender Implicit Association Test (IAT) scores compared to the disparities in participants' review behaviors and decisions.

The IAT Scores range from -2 to 2, where a higher score indicates a stronger inclination to associate females with liberal arts (Gender IAT) or Black with bad (Race IAT). For each participant, we computed the differences in (i) average time spent on applicants of different gender and race groups and (ii) average competitiveness ratings for applicants of different gender and race groups. We conducted linear regression analyses to predict the gender and race IAT scores using these differences. The core assumptions of the linear regression analysis were checked for each model using residual versus fitted value plots and normal Q-Q residual plots, and no significant deviations were observed.

We found **no support for H2.1 and H2.2**: a higher gender IAT score (stronger implicit association of females $\rightarrow$ liberal arts and males $\rightarrow$ science) did not result in participants spending more time on male applicants than female applicants or giving higher ratings to male applicants compared to female/non-binary applicants ($Coef. < 0$). However, we found **support for H2.3 and H2.4:** participants with higher gender IAT scores spent more time on male applicants than non-binary applicants ($Coef. = 0.044$, $p = 0.019$, corresponding to 0.76 min difference between the reviewers with the lowest possible gender IAT score and the highest). Similarly, we observed that participants with higher race IAT scores (stronger implicit association of White $\rightarrow$ good and Black $\rightarrow$ bad) indeed spent more time ($Coef. = 0.096$, $p < 0.001$; corresponding to 1.67 min difference between the reviewers with lowest possible race IAT score and the highest) and assigned higher ratings ($Coef. = 0.08$, $p < 0.001$; corresponding to 16.92 out of 100 difference in competitiveness ratings between reviewers with lowest and highest race IAT score) to White applicants than Black applicants.

7.6.4.3 H3: Effect of the Intervention

We analyzed participants' interactions on the summary page and compared competitiveness ratings and time spent before and after seeing the summary page. Additionally, we summarize qualitative feedback from participants.

Interaction Analysis. We analyzed logs of click and hover interactions to understand how participants used the interface. *Hover* interactions suggest instances where participants explored their ratings and time spent on applicants. 71 participants (95%) initiated 2192 hover events (1500 from the applicant list and 667 from the strip plots). *Clicking* on points in the strip plots or applicant names indicates revisiting the rating page for the corresponding application. There were 80 revisits from 44 (59%) participants and each of these participants revisited on average 2 applicants ($max = 7$, $min = 1$). Most revisits occurred through the applicant list (53), while about 1/3 of the revisits were initiated from the strip plots (26). Notably, there were more interactions on the strip plots representing Time Spent (456 hovers + 14 clicks) than the strip plots representing Competitiveness Ratings (211 hovers + 12 clicks). This could be attributable to participants' expressed surprise on the Time Spent plots as discussed in the following sections. We found 26 changes to applicant ratings from these revisits (19 upgrades, 7 downgrades) in total for 11 applicants from 9 participants.

We observed certain trends in the types of applicants that participants interacted with: in the Time Spent by Gender and Competitiveness Rating by Gender strip plots, participants interacted with more female applicants compared to non-binary and male applicants; and in the Time Spent by Race and Competitiveness Rating by Race plots, participants interacted with more Black applicants compared with White applicants. The detailed interaction counts are summarized in Supplemental Materials. Furthermore, we found that most of the interactions (10/12 clicks, 144/211 hovers) on the Competitiveness Rating plots corresponded to applicants who were rated lower (below the average). Similarly, most of the click interactions (9/12) on the Time Spent by Gender plot corresponded to applicants who received less review time (less than the average).

Change of Time Spent and Competitiveness Ratings. We compared participants' time spent and competitiveness ratings across different gender/race groups before and after interacting with the summary page and found that the changes are marginal at the aggregate level. However, we found that the summary page had a **substantial impact on some individual participants' review behavior and decisions**. For example, before interacting with the summary page, $P_{CTRL}69$ spent on average more time on White applicants than Black applicants ($\Delta_{\text{mean\_time(W-B)}} = 2.245 - 1.843 = 0.402\,\text{min}$) and rated White applicants higher than Black applicants ($\Delta_{\text{mean\_rating(W-B)}} = 78.000 - 72.333 = 5.667$). The participant indicated surprise and disappointment about the behavior—*"I was also surprised, and perhaps disappointed with myself, for giving white people higher scores and more time than black people. I was not conscious that I was doing this and it is good to know."* and made adjustments to their evaluation process. As a result, the disparities in time spent and competitiveness ratings among Black and White applicants decreased ($\Delta_{\text{mean\_time(W-B)}} = 2.408 - 2.292 = 0.115\,\text{min}$, $\Delta_{\text{mean\_rating(W-B)}} = 77.667 - 74.167 = 3.50$) after the participant revisited applicants and adjusted competitiveness ratings.

Similarly, participant $P_{CTRL}40$ initially rated male applicants higher than female applicants on average with a substantial disparity ($\Delta_{\text{mean\_rating(M-F)}} = 83.750 - 64.750 = 19$). In response, the participant indicated *"I think I ultimately revised some answers to ensure no disparities apart from objective criteria, in some cases I simply hadn't remembered correctly where I had ranked other students of similar GPA and class rank so I rectified that."* Consequently, the magnitude of the disparity decreased significantly, and the direction changed to favor female applicants($\Delta_{\text{mean\_rating(M-F)}} = 64.00 - 67.25 = -3.25$).

Qualitative Feedback. *Usefulness.* Participants rated the usefulness of the Time Spent (Fig. 7.19 H) and the Competitiveness Rating (Fig. 7.19 I) strip plots on a 5-point scale (1: Not at all useful, 5: Extremely useful). The mean score was 2.9 (SD = 1.1) for the Time Spent plots and 3.2 (SD = 1.2) for the Competitiveness Rating plots.

Many participants (7 for the Time Spent plots, 15 for the Competitiveness Rating plots) commented the plots helped them **assess fairness and potential bias** (e.g., $P_{CTRL}79$ expressed *"It gave me an idea of how much effort I put into each individual and assured me on how fair I treated each individual's information in arriving at my decision."* Some participants (5) also mentioned that the time spent plots could lead to **adjustment of behavior** (e.g., $P_{CTRL}25$ commented *"I think it could help you reconsider an application if you spent significantly less time on it, and maybe taking another look at an applicant you may have initially dismissed"*). Some participants (3) commented that the Competitiveness Rating plots would be *"more useful if there was a larger data pool"* ($P_{CTRL}68$) suggesting that the visualizations could have more impact in real-world admissions scenarios where the application pool is much larger.

Participants who found the visualizations not useful emphasized that they were fair—their evaluations were based on applicants' merit without considering gender or race. Some participants thought the time spent plots were not useful because the time spent on applications was influenced by different factors (*"I didn't find it particularly useful just because there were other factors at play."* - $P_{CTRL}8$) including the order in which they were reviewed and the content of the application package.

Surprise. Participants rated whether they were surprised by their data shown on the Time Spent plots (Fig. 7.19 H) and the Competitiveness Rating plots (Fig. 7.19 I) on a 5-point scale (1: Not at all surprised, 5: Extremely surprised). In general, participants expressed no/low surprise in their data with a mean score of 1.9 (SD = 1.0) for Time Spent plots and 1.7 (SD = 1.0) for Competitiveness Rating plots. Many participants who were not surprised mentioned that gender and race were not factors that impacted their reviews. In some cases, participants expressed surprise that they rated applicants in a certain group higher (e.g., *"I was surprised that I tended to rate male applicants higher than female applicants."* - $P_{CTRL}44$) or spent more time on a certain group of applicants unconsciously (e.g., *"I wasn't aware that I spent more time reviewing non-binary applicants."* -$P_{CTRL}12$).

7.6.5 Case Study Methodology

We conducted a case study in the Computer Science department at a private university where the system was used for the department's Ph.D. admissions reviews over a time period of roughly two months. This case study allowed us to assess real-world efficacy of RUTABAGA for bias awareness.

7.6.5.1 Participants

We recruited participants among the faculty in the Computer Science Department at a large private university. The admissions committee chair sent out an email to all admissions committee members with an introduction and training materials (documentation and video tutorial) of RUTABAGA. The committee members were recommended (but not required) to use RUTABAGA; they could still elect to use prior application review methods (Sect. 7.6.1.2). The two committee chairs and 11/12 committee members *used the system* in some capacity during the admissions process. We were able to subsequently *interview* the two committee chairs and nine committee members after the admissions process concluded. They had 0 to 14 ($\mu = 4$) years of prior involvement in admissions, offering a diverse range of perspectives. We refer to participants in the "Results" section (Sect. 7.6.6) as P1-13. Among these, P1 - P11 used the system and were interviewed, P12 - P13 were not interviewed, and P9 - P11 only used the system partially. We note that one of the authors was a member of the admissions committee (P3). We include their data in our analysis in order to present a comprehensive and transparent view of the admissions review process. Except for P3 and the two committee chairs, none of the participants had prior experience using the system before the case study.

Given the complexity of deploying new tools in real-world scenarios, our study focused on a single department to ensure a thorough evaluation. Although the sample size is limited, it includes experienced decision-makers directly involved in the admissions process, capturing key decision-making dynamics—with committee chairs overseeing the process and faculty members evaluating applications. The study allowed us to gather detailed feedback on the system's usability, effectiveness, and potential to promote bias awareness. The insights gathered, while specific to a single department, are valuable for understanding the system's potential in similar contexts. We discuss the limitations of relying on a convenience sample and its potential impact on the generalizability of the system design in Sect. 7.6.7. Future work will aim to address these limitations by expanding the participant pool to include a wider range of programs and institutions, ensuring that the system is robust and adaptable to various admissions contexts.

7.6.5.2 Dataset

There were 161 Ph.D. applications in total, and each application was reviewed by at least two committee members. Each application consists of general form field information including research interests, education background, and optional test scores; required document uploads, including personal statement, resume, up to four recommendation letters, and transcripts; and optional files including writing samples and diversity statement. The applications were downloaded from the application portal and loaded to the database for RUTABAGA by the first author before the system was available for reviewers.

7.6.5.3 Procedure

During the first two weeks of the admissions review phase, participants independently reviewed and rated assigned applications. Next, the first author loaded the scores and notes provided in a spreadsheet from faculty members who did not use the system, then enabled the Group Summary Page before the first committee meeting. During the meeting, the Group Summary Page (described in Supplemental Materials) was used to facilitate the discussion via screensharing on Zoom. The committee decided which candidates were not a good fit for the program (assigned to the Reject list) and the rest of the candidates advanced to the Interview round which occurred over the subsequent two-week period. A second committee meeting took place after the interviews where the final admissions decisions were made.

After the admissions cycle concluded, emails were sent to all the committee members to invite them for an interview in order to gather user feedback. All the interviews were conducted via zoom and lasted 30 to 60 min. Upon providing informed consent, the interviews were screen-recorded. After gathering background information, the interviewer asked the participant to login to the system and share their screen to facilitate a walk-through of the system, discuss their experience, and provide suggestions for improvements. Following the interview, the participants were asked to complete a post-study questionnaire, consisting of questions about the usability of the system (Sect. 7.6.6.4).

7.6.5.4 Analysis and Coding

The first author first used an audio-to-text tool to transcribe the interview recordings, and then manually refined the transcripts. The research team used qualitative data analysis methods [82] to analyze the interview transcripts. Specifically, thematic analysis was conducted on the interviews through inductive coding. Two authors independently coded two interview transcripts and discussed to develop a codebook. After refining coding definitions together, the first author coded the remaining transcripts. The final codebook contains 36 codes in nine categories including System Usage, Review Strategy, Awareness, and others, the details of which are included in Supplementary Materials.

7.6.6 Case Study Results

We organize results according to high-level themes of the case study, including increasing participants' **awareness** of their review process (Sect. 7.6.6.1), associated **behavioral** changes (Sect. 7.6.6.2) and changes in **decisions** (Sect. 7.6.6.3), consistent with [119]. System **usability** scores and qualitative feedback are presented in Sect. 7.6.6.4.

While 11 reviewers used the Rating Page, only five of them used the Summary Page, although all provided feedback after interacting with the view during the interview. All of the reviewers interacted with the Group Summary Page directly or indirectly (through screen-share from the committee chair) during the committee meetings.

7.6.6.1 Awareness

We use the term **awareness** to refer to insights gained from reflecting on one's review process via interaction history on the applications. The findings in this section were derived from the interviews we conducted with the participants.

Process Awareness. Three participants found that features in the Summary Page helped them to systematically reflect on fairness in their time spent across applications and application components (*"I was able to look at the amount of time that I spent over their different documents just to make sure I didn't miss anything."*—P8) and adjust their review behaviors when they identified undesired behaviors. For instance, the time spent distribution chart helped P8 be **aware** that (*"I hadn't really spent time on their writing sample"*), led them to different **behaviors** (*"So I went back and I had the chance to read over it"*), and make changes to some **decisions** (*"it was a good chance to revise what I had scored for their research preparedness"*). The scatterplot helped P3 identify outliers in the time they spent on applications (*"I just didn't spend that much time on someone"*) and led to different

behaviors (*"I would try to go back and (...) just spend a little bit more time looking at them and see if I missed something"*). P5 commented, *"the folks that I may have not spent as much time on, I'd want to know why."*

Outcome Awareness. In addition to assessing internal consistency of time spent, three participants also used the Summary Page to check internal consistency in their ratings (*"I check the different distributions just to make sure that I was sort of consistent in giving my overall ranking."*—P8. P8 tried to self-calibrate on the ratings (*"we had the chance to go back to our reviews, compare the ones that we had scored previously, and change our rating"*). The scatterplot helped P3 identify outliers in the ratings (*"I identified outliers like someone that I rated as having high research preparedness but I did not rate them overall very highly."*) and revisited the applications (change in **behaviors**)—*"I would go back and look at their applications again"*. The committee chairs planned to *"take a look at the scatterplot to evaluate our process overall"* in the Group Summary Page.

Fairness/Bias. Four participants found the system useful in terms of increasing awareness about procedural fairness. P5 liked that the profile view allows hiding attributes (*"I really like that... I basically turned off anything that I felt might bias my decision"*). P8 was interested in seeing *"was there really some sort of unintentional way of aspects that influence my decision."* By looking at the scatterplot in the Summary Page with different combinations of X- and Y-axis attributes, the participant found that *"there's no bias toward any gender or race in my decision. That was good for me to know."*

During the interview, participants who did not actively use the Summary Page during the admission process tried to interact with the interface and found the scatterplot *"is showing how I have reviewed people (...) if I have some gender bias or race bias."* (P4), and the time spent chart could answer the question *"are you spending the right amount of time or at least enough time on all the different applicants?"* (P2), indicating the system has the potential to increase reviewers' awareness of bias during the review process, even if they did not initially use the system as such.

7.6.6.2 Interaction Analysis

Interactions with Individual Summary. The system logged users' interactions with the Summary Page, including hovering on the points on the scatterplot (Fig. 7.19F) (to see the applicant's information), clicking on the points (to revisit the Rating Page), and clicking on the Overall Recommendation radio buttons (to modify overall recommendation). We analyzed this interaction data to understand if and how participants used this part of the interface.

We found that reviewers visited the Summary Page at different phases of the admission process, i.e., during the individual review phase and during the group meeting. Of the 11 reviewers who used the Rating page, five visited the Summary Page. Among these, three actively interacted with the page during both phases and the other two reviewers interacted

Table 7.2 Participants' interactions with the summary page. Numbers in parentheses represent interactions during the independent review phase

	# Hover	# Revisit	# Changes in recommendation
P1	22 (5)	4 (1)	0 (1)
P3	4 (13)	1 (3)	0 (2)
P5	5	0	0
P8	10 (27)	8 (3)	0 (3)
P12	11	1	0

with the page only during the group meeting. Table 7.2 shows the number of distinct applicants these five reviewers interacted with in the two phases. Although the numbers are too small to draw generalizable conclusions, we further looked into what type of applicants reviewers tended to hover and click on. During the independent review phase, P8 hovered on almost all the applicants they reviewed, revisited applicants with low overall recommendation, and rated the applicants higher after revisits. P1 mostly hovered on applicants who they rated as Not Competitive and rated one of them higher afterward. P3 hovered on and revisited more competitive applicants and both downgraded and upgraded some applicants.

Time Spent. One-way ANOVA shows that at the aggregate level, there were no significant differences in the time spent across gender and race. To account for differences in application characteristics, we normalized the time spent by incorporating a complexity score for each application. This complexity score was calculated as the sum of the normalized portfolio length and normalized word count. By normalizing time spent, we ensure that comparisons reflect differences in reviewer behavior rather than variability in the complexity of the applications themselves. However, similar to our findings from the controlled study, we observed individual differences on how reviewers spent time across gender. As shown in Fig. 7.21, some reviewers (e.g., P3 and P4) spent more time on Female applicants on average, while others (e.g., P2 and P12) spent more time on Male applicants. Among the 10 participants in Fig. 7.21, two identified as female (P3 and P5), while eight identified as male. Of the female participants, one spent more time reviewing female applicants, while the other spent more time reviewing male applicants. The review behavior among the male participants was also variable. Some (e.g., P4 and P6) spent more time reviewing female applicants, some (e.g., P2 and P7) spent more time reviewing male applicants. Given the small sample size, it is not feasible to identify a consistent trend in review behavior linked to reviewer gender.

Those who interacted with the awareness features (P1, P3, P8) tended to have relatively small disparities in time spent between male and female applicants (P1, P8) or, in some cases, reversed the trend to spend more time on female applicants (P3). We note that although many comparisons did not result in statistically significant differences at the group level, the

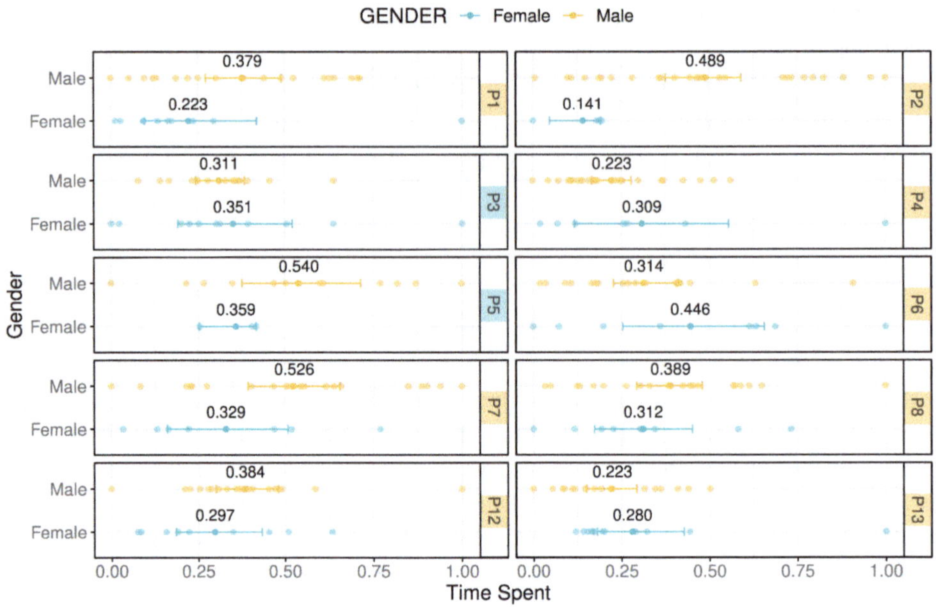

Fig. 7.21 Strip plots comparing time spent (normalized) on applicants by each reviewer grouped by applicant gender with bootstrapped 95% CI. The mean value is annotated for each gender and each participant. The background color of each participant label represents the reviewer's gender: blue for female participants and yellow for male participants. Three participants' data were excluded due to limited usage of the system for application review

analyses of individual reviewer behavior can be cause for further scrutiny. We discuss this, along with potential explanations of these trends in Sect. 7.6.7.

7.6.6.3 Decisions

The distribution of admitted, waitlisted and rejected applicants by gender and race aligns to the underlying distribution of the candidate pool, i.e., there was no clear favor of a certain group when making decisions (Fig. 7.22), confirmed by a χ^2 test ($p > 0.1$). As described in Sect. 7.6.6.2, reviewers' individual recommendations for some applicants were altered after interacting with the Summary page. There were four upgrades and two downgrades from 3 reviewers on 6 applicants. Two of the applicants whose ratings were upgraded were admitted while the two downgraded applicants were rejected, suggesting that the intervention could have had a critical impact on the outcomes for these applicants, although we cannot say with certainty what the outcome would have been for these applicants otherwise.

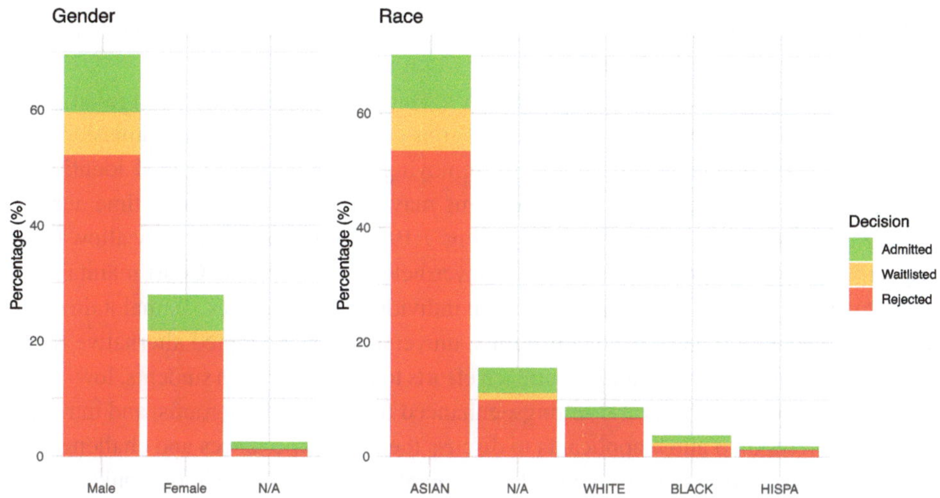

Fig. 7.22 The distribution of admissions decisions across applicant gender and race

7.6.6.4 System Usability

In the post-study questionnaire, participants rated their overall experience with the system as well as impressions on different features (details are attached in Supplemental Materials). Results of the questionnaire and qualitative feedback from the interviews indicate that participants' overall experience with RUTABAGA was positive.

Participants' average system usability scale (SUS) score is "Good" (score above 68) at 78.3 [18]. Visualization components of the system (the Scatterplot and the Time Spent Distribution) each received divergent usefulness scores ranging from 1 (low) to 5 (high). The mean score was 3.4 ($SD = 1.1$) for the Scatterplot and 3.5 ($SD = 1.1$) for the Time Spent Distribution. P11 commented that the scatterplot can be useful for identifying patterns and suggested providing guidance on how to perform analyses with it. Participants who rated the Scatterplot relatively low commented that *"faculty are (...) very busy... So we don't necessarily spend maybe enough time being self critical"* and suggested that the analysis can be done by administrators (P2). We discuss future work on incorporating these suggestions in Sect. 7.6.7.

7.6.7 System Implications

Implications of Affirmative Action. Affirmative action was a policy that arose in the United States in the 1960s to reinforce the Civil Rights Act of 1964 aimed at eliminating discrimination [3]. It intended to create opportunities for traditionally disadvantaged and underrepresented groups in workplace and educational settings and thus, alongside standard evaluation criteria, involved explicit consideration of additional factors such as race and gender in application review processes [45].

The 2023 Supreme Court decisions (SFFA v. Harvard and SFFA v. UNC) ruled against considering race as an explicit factor of admissions decisions [113]. Many admissions committees are operationalizing the ruling by removing direct access to racial identity data *during* the admissions process. Thus programs may choose to restrict real-time access to visualizations shown in the Summary page (Fig. 7.19, middle) to instead only allow for retrospective analysis to ensure compliance. Nevertheless, the Supreme Court maintains that race may be considered insofar as it affects an individual's experiences. To maintain diverse student bodies within these legal constraints, universities have adopted alternative recruiting strategies. Many have expanded outreach efforts to first-generation students, low-income students, and students from rural settings, enhanced financial aid programs, and introduced new essay prompts to allow applicants to discuss their life experiences and challenges [23, 60]. Despite these efforts, the ruling has led to notable changes in the demographic composition of new classes at many top institutions, with black student enrollment declining [23, 60]. Navigating the legality of ensuring diversity, equity, and inclusion in university admissions remains an ongoing challenge in the wake of SFFA v. Harvard and SFFA v. UNC rulings. As a result, we maintain that the importance of continued research on ensuring fair review processes has never been more critical.

In the future, we can explore methods to assist admissions committees in meaningfully accounting for the differences in opportunities and privileges among diverse applicants. For instance, new system features could be developed to help reviewers identify various forms of adversity faced by applicants and design targeted interventions to ensure that these students are not overlooked in the admissions process.

Is Time Spent a Good Proxy for Bias? More time spent on an applicant does not necessarily reflect a negative bias. If a reviewer spends less time on a specific applicant, it could be due to unconscious bias, but it could also be due to other benign reasons. For instance, P11 noted *"I just know this candidate and I wrote her letter, and so I'm looking very little at her."* Other reviewers observed that *"just like in a conference reviewing setting, there are some manuscripts that are clear accepts and there's some manuscripts that are clear rejects."*—P2. Other factors also influenced time spent such as a reviewer's familiarity with transcripts from foreign institutions (P6), general readability of other application components (P6), or varying lengths of documents like recommendation letters (P5). Reviewers tended to agree that *"most of the time is being spent on the murky middle"*—P11. Participants from the controlled study also indicated that they tended to naturally spend more time on the first few applicants when getting familiar with the task.

Thus time spent on applications is an inherently noisy metric sensitive to several factors, rather than a definitive measure of bias. However, results of our controlled study in Sect. 5.6.2.2 indicate that **in spite of these sources of noise, time spent can nonetheless provide a meaningful signal** that illustrates an observable effect of implicit bias on review behaviors. In addition, similar to the stated goals for Wall et al.'s bias metrics [121] and consistent with the goals of reflective design [100], our aim in RUTABAGA is to promote *reflection* on potential biases. Thus while time spent is an imperfect proxy for bias, its representation in RUTABAGA can cause reviewers to more carefully reflect on their review process nonetheless. Future work can explore variability in application characteristics, such as readability, to better understand the factors influencing time spent on applications. Incorporating these measures could enhance the interpretation of time spent as a signal for bias and provide deeper insights into review behaviors.

Design Implications. While most of the participants from the controlled study interacted with the features intended for reflection to some extent (Sect. 7.6.4.3), only part of the committee used the features in the case study (Sect. 7.6.6). Despite close collaboration with committee chairs during the system's formative design (Sect. 7.5.2), many committee members nonetheless found the Summary and Group Summary pages to be visually overwhelming and chose not to engage (*"It looks scary to me ... I felt a little bit overwhelmed by what was going on"* -P11). Given our observations from the post-study interviews in the case study, we iterated on the system design to enhance user engagement by making the system more intuitive and accessible. The changes include: (1) adding an embedded Help Page in the system for a more convenient on-boarding, (2) providing an explanation of the Summary page when the user first interacts with the page, (3) adding a text summary for each of the visualizations to reduce cognitive load, and (4) adding pre-defined box-plots to show the user's time spent across different gender/race groups. We observed increased usage of the system in the subsequent admission cycles (27 admissions committee members engaged with the Summary page).

Based on reflection of lessons learned from our studies, we summarize some design considerations for future work creating visual tools for admissions review.

- Augmenting visualizations with text could ease cognitive load. As noted by participants from the controlled study, text summaries of the visualizations (which were absent in the case study) *"helped to ease the processing of the data"*—P_{CTRL}43.
- Providing pre-defined analyses could complement self-exploration. In the case study, participants suggested predefined analyses, e.g., committee chairs configuring views of essential bias analyses such as gender distribution of admitted applicants or time spent by applicant gender individual committee members can start with.
- Providing sufficient onboarding is critical for system adoption. We provided a tutorial video demonstrating the features of the system in the case study. However, some participants didn't know how to use it and *"needed a better training"* (P7). Integrated explanatory features such as clickthrough tutorials, embedded videos, and help pages could increase learnability.
- Exploring the balance of mixed-initiative user interfaces [66, 115] could be a promising direction. The system could gently nudge [108] users to interact with reflective views

or, more aggressively in cases of strict procedural goals, require it, e.g., using pop-up notifications that cannot be dismissed prior to engagement with the analysis.

Impact of the Intervention on Individuals. We observed from the studies that (i) individual behaviors can be highly variable, and (ii) the impact of the awareness features of RUTABAGA is likewise variable. The efficacy of techniques used in RUTABAGA is likely sensitive to reviewers' conscientiousness, time constraints, and other individual factors; some individuals may be resistant to any form of intervention in this setting. However, we are optimistic that our analysis of individual reviewers revealed that RUTABAGA's techniques have the potential to be transformational for some reviewers' processes. The system has demonstrated the potential to cause reflection and subsequent changes in behaviors and decisions for some individuals who are motivated to assess and rectify behavioral disparities.

Generalizability and Scalability. The current system is designed based on one program's Ph.D. admissions process. However, the admissions process can vary widely across different institutions and programs at the graduate and undergraduate levels. For example, Master's admissions processes often prioritize academic performance, such as GPA and standardized test scores, while non-academic factors such as personal statements and letters of recommendation play a less significant role in the review [124]. Undergraduate admissions, on the other hand, typically involve larger applicant pools and often rely on a broader range of holistic evaluation criteria that include factors such as extracurricular activities, socioeconomic status, and academic records [31]. Nevertheless, our design could be adapted to support holistic review processes similar to ours. The specific features like the rating criteria and scales, plot configurations, etc. in the system can be customized for different program needs. However, further work is needed to assess its applicability to programs deviating substantially from ours, e.g., in size of application pool, distribution of reviewer responsibilities, etc. We could envision additional features that may be helpful in the case of large applicant pools. For instance, boxplots or density plots can be used instead of strip plots to present high-level summaries with outliers displayed. Users can observe aggregated trends first and dive into individual-level data as desired. For institutions already employing customized tools (e.g., Slate [102]) for admissions review, capturing user interactions and providing visualizations could be integrated into existing tools by closely working with the tool designers. Seamless integration, however, can be challenging and needs to be explored in future work.

Ethical Considerations in Real-World Deployment. Deploying the system in a real-world admissions context introduces ethical considerations, particularly given its potential to influence applicant outcomes. The deployment of RUTABAGA was guided by the department's goal of improving fairness in the admission process. The system was designed to address the inherent limitations in human decision-making by identifying potential biases and inconsistencies. Its role was not to replace human reviewers, but rather to act as an assistive tool that could identify issues that might otherwise have gone unnoticed. To mitigate potential concerns, admissions committee members were fully informed about the purpose and capabilities of the system, and the use of the system was entirely optional for reviewers.

Limitations. One limitation of our work is that participants in the controlled study were not screened for admissions training. As we discussed in Sect. 7.6.3.1, it is challenging to recruit a substantial sample size of admissions officers, and instead we recruited participants based on a minimum education level of a bachelor's degree as inclusion criteria. Admissions reviewers typically receive diversity, equity, and inclusion (DEI) training, which may influence their decision-making differently from untrained participants. Nevertheless, our findings may be comparable to what could be expected in general workplace application reviews, where the background and training of reviewers tend to be less standardized and more diverse than in Ph.D. admissions [122]. Nevertheless, even in our real-world deployment involving trained Ph.D. admissions reviewers, we observed high variability in review behaviors, underscoring the complexity of decision-making in these contexts. This variability suggests that individual differences play a significant role in reviewer behaviors, regardless of training or experience. However, we cannot speculate on how the findings of this controlled study would generalize to larger sample sizes or other domains without further empirical investigation.

Another limitation of the control study is the limited set of gender and racial categories used. Although our goal was to obtain comparable results with previous work [15] and to create as diverse a range of applicant profiles as possible, we were constrained by the need to limit the study duration to prevent participant fatigue. The exclusion of other racial groups, such as Hispanic or Asian applicants, constrains our ability to explore the broader dynamics of bias and fairness. Different racial categories may reveal unique patterns of reviewer behavior or bias that were not captured here. Future research should incorporate a more comprehensive range of racial categories to better understand how biases operate across diverse populations.

Furthermore, the interviews conducted in the case study may be subject to response bias [46], where participants may provide feedback they perceive as desirable and favorable by the researchers.

References

1. https://github.com/nl4dv/nl4dv/blob/master/examples/assets/data/movies-w-year.csv Accessed 2021-03-29.
2. Chadia Abras, Diane Maloney-Krichmar, and Jenny Preece. "User-Centered Design". In: *Bainbridge, W. Encyclopedia of Human-Computer Interaction. Thousand Oaks: Sage Publications* 37.4 (2004), pp. 445–456.
3. Civil Rights Act. "Civil Rights Act of 1964". In: *Title VII, Equal Employment Opportunities* (1964).

4. Piotr D Adamczyk and Brian P Bailey. "If not now, when?: the effects of interruption at different moments within task execution". In: *Proceedings of the SIGCHI conference on Human factors in computing systems*. ACM. 2004, pp. 271–278.
5. Philip Adler et al. "Auditing black-box models for indirect influence". In: *Knowledge and Information Systems* 54 (2018), pp. 95–122.
6. Yongsu Ahn and Yu-Ru Lin. "FairSight: Visual Analytics for Fairness in Decision Making". In: *IEEE transactions on visualization and computer graphics* PP (Aug. 2019). DOI: https://doi.org/10.1109/TVCG.2019.2934262.
7. John H Aldrich et al. *Why parties?: The origin and transformation of political parties in America*. University of Chicago Press, 1995.
8. Micah Allen et al. "Raincloud plots: a multi-platform tool for robust data visualization". In: *Wellcome open research* 4 (2019).
9. Julia Angwin et al. "Machine Bias". In: *ProPublica* (2016).
10. Hal R Arkes. "Costs and benefits of judgment errors: Implications for debiasing." In: *Psychological bulletin* 110.3 (1991), p. 486.
11. Karthik Badam et al. "Supporting Team-First Visual Analytics through Group Activity Representations". In: *Graphics Interface* (2017).
12. Donald Baker et al. "Providing customized process and situation awareness in the collaboration management infrastructure". In: *Cooperative Information Systems, 1999. CoopIS'99. Proceedings. 1999 IFCIS International Conference on*. IEEE. 1999, pp. 79–91.
13. Aruna D Balakrishnan, Susan R Fussell, and Sara Kiesler. "Do visualizations improve synchronous remote collaboration?" In: *Proceedings of the SIGCHI Conference on Human Factors in Computing Systems*. ACM. 2008, pp. 1227–1236.
14. Rachel Bellamy et al. "AI Fairness 360: An Extensible Toolkit for Detecting and Mitigating Algorithmic Bias". In: *IBM Journal of Research and Development* PP (Sept. 2019), pp. 1–1. DOI: https://doi.org/10.1147/JRD.2019.2942287.
15. Marianne Bertrand and Sendhil Mullainathan. "Are Emily and Greg more employable than Lakisha and Jamal? A field experiment on labor market discrimination". In: *American economic review* 94.4 (2004), pp. 991–1013.
16. Alex Beutel et al. "Beyond globally optimal: Focused learning for improved recommendations". In: *International Conference on World Wide Web*. 2017, pp. 203–212.
17. Fatma Bouali, Abdelheq Guettala, and Gilles Venturini. "VizAssist: an interactive user assistant for visual data mining". In: *The Visual Computer* 32.11 (2016), pp. 1447–1463.
18. John Brooke. "SUS: A Retrospective". In: *Journal of usability studies* 8.2 (2013), pp. 29–40.
19. Michael Brooks et al. "FeatureInsight: Visual support for error-driven feature ideation in text classification". In: *IEEE Conference on Visual Analytics Science and Technology*. IEEE. 2015, pp. 105–112.
20. Eli T Brown et al. "Finding waldo: Learning about users from their interactions". In: *IEEE Transactions on visualization and computer graphics* 20.12 (2014), pp. 1663–1672. DOI: https://doi.org/10.1109/TVCG.2014.2346575.
21. Eli T. Brown et al. "Dis-Function: Learning Distance Functions Interactively". In: *IEEE Conference on Visual Analytics Science and Technology (VAST)* (2012), pp. 83–92.
22. Daniel Burgstahler et al. "Switching push and pull: An energy efficient notification approach". In: *Mobile Services (MS), 2014 IEEE International Conference on*. IEEE. 2014, pp. 68–75.
23. Hilary Burns and Neena Hagen. " 'It is heartbreaking': Black first-year enrollment plunges at top colleges, post-affirmative action". In: *The Boston Globe* (2024). URL: https://www.bostonglobe.com/2024/09/25/metro/affirmative-action-colleges-black-enrollment/.
24. Angel Cabrera et al. "FAIRVIS: Visual Analytics for Discovering Intersectional Bias in Machine Learning". In: Oct. 2019, pp. 46–56. DOI: https://doi.org/10.1109/VAST47406.2019.8986948.

25. Toon Calders and Sicco Verwer. "Three naive bayes approaches for discrimination-free classification". In: *Data mining and knowledge discovery* 21 (2010), pp. 277–292.
26. Flavio Calmon et al. "Optimized Data Pre-Processing for Discrimination Prevention". In: (Apr. 2017).
27. L Elisa Celis et al. "Classification with fairness constraints: A meta-algorithm with provable guarantees". In: *Proceedings of the conference on fairness, accountability, and transparency*. 2019, pp. 319–328.
28. Davide Ceneda et al. "Characterizing guidance in visual analytics". In: *IEEE Transactions on Visualization and Computer Graphics* 23.1 (2017), pp. 111–120.
29. Isaac Cho et al. "The Anchoring Effect in Decision-Making with Visual Analytics". In: *IEEE Conference on Visual Analytics Science and Technology (VAST)* (2017).
30. William S Cleveland and Robert McGill. "Graphical perception: Theory, experimentation, and application to the development of graphical methods". In: *Journal of the American statistical association* 79.387 (1984), pp. 531–554.
31. Arthur L Coleman and Jamie Lewis Keith. "Understanding Holistic Review in Higher Education Admissions". In: *New York: College Board* (2018).
32. Mary Czerwinski, Edward Cutrell, and Eric Horvitz. "Instant messaging: Effects of relevance and timing". In: *People and computers XIV: Proceedings of HCI*. Vol. 2. 2000, pp. 71–76.
33. Yanan Da et al. "rutaBAGA: A Visualization Approach for Bias Awareness in University Admissions". In: *Journal of Visualization and Interaction* (2025).
34. Evanthia Dimara, Anastasia Bezerianos, and Pierre Dragicevic. "The attraction effect in information visualization". In: *IEEE transactions on visualization and computer graphics* 23.1 (2017), pp. 471–480.
35. Evanthia Dimara et al. "Mitigating the attraction effect with visualizations". In: *IEEE transactions on visualization and computer graphics* 25.1 (2019), pp. 850–860.
36. Wenwen Dou et al. "Recovering Reasoning Process From User Interactions". In: *IEEE Computer Graphics & Applications* May/June.March (2009), pp. 52–61. URL: https://doi.org/10.1.1.157.407%7B%5C&%7Drep=rep1%7B%5C&%7Dtype=pdf.
37. Pierre Dragicevic. "Fair statistical communication in HCI". In: *Modern statistical methods for HCI*. Springer, 2016, pp. 291–330.
38. Alex Endert, Patrick Fiaux, and Chris North. "Semantic interaction for visual text analytics". In: *Proceedings of the 2012 ACM annual conference on Human Factors in Computing Systems - CHI '12* (2012), pp. 473–482. URL: http://dl.acm.org/citation.cfm?doid=2207676.2207741.
39. Mica R Endsley. "Toward a theory of situation awareness in dynamic systems". In: *Human factors* 37.1 (1995), pp. 32–64.
40. Mussweiler Englich and T Mussweiler. "Anchoring effect". In: *Cognitive Illusions: Intriguing Phenomena in Judgement, Thinking, and Memory* (2016), p. 223.
41. Mi Feng, Evan Peck, and Lane Harrison. "Patterns and Pace: Quantifying Diverse Exploration Behavior with Visualizations on the Web". In: *IEEE transactions on visualization and computer graphics* 25.1 (2019), pp. 501–511.
42. Mi Feng et al. "Hindsight: Encouraging exploration through direct encoding of personal interaction history". In: *IEEE transactions on visualization and computer graphics* 23.1 (2016), pp. 351–360.
43. Stephen Few. *Now you see it: simple visualization techniques for quantitative analysis*. Analytics Press, 2009.
44. Sorelle Friedler, Carlos Scheidegger, and Suresh Venkatasubramanian. "Certifying and Removing Disparate Impact". In: (Dec. 2014). DOI: https://doi.org/10.1145/2783258.2783311.
45. Robert Fullinwider. "Affirmative Action". In: *The Stanford Encyclopedia of Philosophy*. Ed. by Edward N. Zalta. Summer 2018. Metaphysics Research Lab, Stanford University, 2018.

46. Adrian Furnham. "Response Bias, Social Desirability and Dissimulation". In: *Personality and individual differences* 7.3 (1986), pp. 385–400. DOI: https://doi.org/10.1016/0191-8869(86)90014-0. URL: https://doi.org/10.1016/0191-8869(86)90014-0.
47. *Georgia General Assembly*. https://www.legis.ga.gov Accessed 2021-03-04.
48. *Georgia State Senate Committees*. http://www.senate.ga.gov/committees/en-US/Home.aspx Accessed 2020-04-02.
49. Gerd Gigerenzer. "Fast and frugal heuristics: The tools of bounded rationality". In: *Blackwell handbook of judgment and decision making* 62 (2004), p. 88.
50. Stefan Gladisch, Heidrun Schumann, and Christian Tominski. "Navigation recommendations for exploring hierarchical graphs". In: *International Symposium on Visual Computing*. Springer. 2013, pp. 36–47.
51. David Gotz, Shun Sun, and Nan Cao. "Adaptive contextualization: Combating bias during high-dimensional visualization and data selection". In: *Proceedings of the 21st International Conference on Intelligent User Interfaces*. ACM. 2016, pp. 85–95.
52. Vincent Grari, Sylvain Lamprier, and Marcin Detyniecki. "Fairness without the sensitive attribute via causal variational autoencoder". In: *arXiv preprint* arXiv:2109.04999 (2021).
53. Anthony G Greenwald and Mahzarin R Banaji. "Implicit Social Cognition: Attitudes, Self-Esteem, and Stereotypes". In: *Psychological review* 102.1 (1995), p. 4. DOI: https://doi.org/10.1037//0033-295x.102.1.4. URL: https://doi.org/10.1037//0033-295x.102.1.4.
54. Anthony G Greenwald and Linda Hamilton Krieger. "Implicit bias: Scientific foundations". In: *California Law Review* 94.4 (2006), pp. 945–967.
55. Anthony G Greenwald, Debbie E McGhee, and Jordan LK Schwartz. "Measuring individual differences in implicit cognition: the implicit association test." In: *Journal of personality and social psychology* 74.6 (1998), p. 1464.
56. Carl Gutwin and Saul Greenberg. "Design for individuals, design for groups: tradeoffs between power and workspace awareness". In: *Proceedings of the 1998 ACM conference on Computer supported cooperative work*. ACM. 1998, pp. 207–216.
57. Carl Gutwin, Mark Roseman, and Saul Greenberg. "A usability study of awareness widgets in a shared workspace groupware system". In: *Proceedings of the 1996 ACM conference on Computer supported cooperative work*. ACM. 1996, pp. 258–267.
58. Diane F Halpern. "Teaching critical thinking for transfer across domains: Disposition, skills, structure training, and metacognitive monitoring." In: *American psychologist* 53.4 (1998), p. 449.
59. Moritz Hardt, Eric Price, Nati Srebro, et al. "Equality of opportunity in supervised learning". In: *Advances in Neural Information Processing Systems*. 2016, pp. 3315–3323.
60. Anemona Hartocollis. "Harvard's Black Student Enrollment Dips After Affirmative Action Ends". In: *The New York Times* (2024). URL: https://www.nytimes.com/2024/09/11/us/harvard-affirmative-action-diversity-admissions.html.
61. Jeffrey Heer et al. "Graphical histories for visualization: Supporting analysis, communication, and evaluation". In: *IEEE transactions on visualization and computer graphics* 14.6 (2008).
62. Richards J Heuer Jr, Richards J Heuer, and Randolph H Pherson. *Structured analytic techniques for intelligence analysis*. Cq Press, 2010.
63. Richards J. Heuer Jr. *Psychology of Intelligence Analysis*. Washington, DC, 1999.
64. Robin M Hogarth. "A note on aggregating opinions". In: *Organizational behavior and human performance* 21.1 (1978), pp. 40–46.
65. Eric Horvitz. "Principles of Mixed-Initiative User Interfaces". In: *Proceedings of the SIGCHI conference on Human Factors in Computing Systems* May (1999), pp. 159–166. DOI: https://doi.org/10.1145/302979.303030.

66. Eric Horvitz. "Principles of Mixed-Initiative User Interfaces". In: *Proceedings of the SIGCHI conference on Human Factors in Computing Systems*. 1999, pp. 159–166. DOI: https://doi.org/10.1145/302979.303030. URL: https://doi.org/10.1145/302979.303030.
67. TJ Jankun-Kelly. "Using visualization process graphs to improve visualization exploration". In: *International Provenance and Annotation Workshop*. Springer. 2008, pp. 78–91.
68. TJ Jankun-Kelly and Kwan-Liu Ma. "A spreadsheet interface for visualization exploration". In: *Proceedings of the Conference on Visualization'00*. IEEE Computer Society Press. 2000, pp. 69–76.
69. Charles E Jordan, Stanley J Clark, and Marilyn A Waldron. "Gender bias and compensation in the executive suite of the Fortune 100". In: *Journal of Organizational Culture, Communications and Conflict* 11.1 (2007), p. 19.
70. Faisal Kamiran and Toon Calders. "Data Pre-Processing Techniques for Classification without Discrimination". In: *Knowledge and Information Systems* 33 (Oct. 2011). DOI: https://doi.org/10.1007/s10115-011-0463-8.
71. Faisal Kamiran, Asim Karim, and Xiangliang Zhang. "Decision theory for discrimination-aware classification". In: *IEEE International Conference on Data Mining*. IEEE. 2012, pp. 924–929.
72. Toshihiro Kamishima et al. "Fairness-aware classifier with prejudice remover regularizer". In: *Joint European Conference on Machine Learning and Knowledge Discovery in Databases*. 2012.
73. Josua Krause, Adam Perer, and Enrico Bertini. "INFUSE: interactive feature selection for predictive modeling of high dimensional data". In: *IEEE TVCG* 20.12 (2014), pp. 1614–1623.
74. Donald R Kretz. "Experimentally Evaluating Bias-Reducing Visual Analytics Techniques in Intelligence Analysis". In: *Cognitive Biases in Visualizations*. Springer, 2018, pp. 111–135.
75. Po-Ming Law and Rahul C Basole. "Designing Breadth-Oriented Data Exploration for Mitigating Cognitive Biases". In: *Cognitive Biases in Visualizations*. Springer, 2018, pp. 149–159.
76. Sukwon Lee, Sung-Hee Kim, and Bum Chul Kwon. "Vlat: Development of a visualization literacy assessment test". In: *IEEE transactions on visualization and computer graphics* 23.1 (2017), pp. 551–560. DOI: https://doi.org/10.1109/TVCG.2016.2598920.
77. Yanhui Li et al. "Training data debugging for the fairness of machine learning software". In: *Proceedings of the 44th International Conference on Software Engineering*. 2022, pp. 2215–2227.
78. Albert E Mannes, Richard P Larrick, and Jack B Soll. "The social psychology of the wisdom of crowds." In: (2012).
79. Jean-Philippe Martin-Flatin. "Push vs. pull in web-based network management". In: *Integrated Network Management, 1999. Distributed Management for the Networked Millennium. Proceedings of the Sixth IFIP/IEEE International Symposium on*. IEEE. 1999, pp. 3–18.
80. Barbara J McNeil et al. "On the Elicitation of Preferences for Alternative Therapies". In: *New England Journal of Medicine* 306.21 (1982), pp. 1259–1262.
81. *Membership of the 115th Congress: A Profile*. https://www.senate.gov/CRSpubs/b8f6293e-c235-40fd-b895-6474d0f8e809.pdf Accessed 2019-07-25. 2018.
82. Matthew B Miles, A Michael Huberman, and Johnny Saldaña. *Qualitative Data Analysis: A Methods Sourcebook*. Sage publications, 2018.
83. Thomas Mussweiler, Fritz Strack, and Tim Pfeiffer. "Overcoming the inevitable anchoring effect: Considering the opposite compensates for selective accessibility". In: *Personality and Social Psychology Bulletin* 26.9 (2000), pp. 1142–1150.
84. Arpit Narechania et al. "Lumos: Increasing awareness of analytic behavior during visual data analysis". In: *arXiv preprint* arXiv:2108.02909 (2021).
85. Arpit Narechania et al. "ProvenanceWidgets: A library of ui control elements to track and dynamically overlay analytic provenance". In: *IEEE Transactions on Visualization and Computer Graphics* (2024).

86. Raymond S Nickerson. "Confirmation bias: A ubiquitous phenomenon in many guises". In: *Review of general psychology* 2.2 (1998), pp. 175–220.
87. Chris North et al. "Analytic Provenance: Process + Interaction + Insight". In: *29th Annual CHI Conference on Human Factors in Computing Systems, CHI 2011* (2011), pp. 33–36. DOI: https://doi.org/10.1145/1979742.1979570. URL: http://www.scopus.com/inward/record.url?eid=2-s2.0-79957960481%7B%5C&%7DpartnerID=40%7B%5C&%7Dmd5=6e451badea5da358cb4d67c3595dfdf3.
88. Chris North et al. "Analytic Provenance: Process+interaction+insight". In: *CHI'11 Extended Abstracts on Human Factors in Computing Systems*. 2011, pp. 33–36. DOI: https://doi.org/10.1145/1979742.1979570. URL: http://dx.doi.org/10.1145/1979742.1979570.
89. Luca Oneto et al. "Taking advantage of multitask learning for fair classification". In: *Proceedings of the 2019 AAAI/ACM Conference on AI, Ethics, and Society*. 2019, pp. 227–237.
90. Dražen Prelec, H Sebastian Seung, and John McCoy. "A solution to the single-question crowd wisdom problem". In: *Nature* 541.7638 (2017), p. 532.
91. Associated Press. "Federal Judge Strikes Down Georgia Abortion Restrictions". In: *Georgia Public Broadcast (GPB)* (2020). URL: https://www.gpb.org/news/2020/07/13/federal-judge-strikes-down-georgia-abortion-restrictions.
92. *Project Implicit*. https://implicit.harvard.edu/implicit/index.jsp. Accessed: 2023-04-30.
93. *Random Name Generator*. https://github.com/treyhunner/names Accessed 2019-07-25. 2014.
94. Megan M Ringel, Cristian G Rodriguez, and Peter H Ditto. "What is right is right: A three-part account of how ideology shapes factual belief". In: *Belief systems and the perception of reality. Oxon: Routledge* (2019).
95. Vanessa Romo. "Georgia's Governor Signs 'Fetal Heartbeat' Abortion Law". In: *NPR* (2019). URL: https://www.npr.org/2019/05/07/721028329/georgias-governor-signs-fetal-heartbeat-law.
96. Bahador Saket, Alex Endert, and Cagatay Demiralp. "Task-based effectiveness of basic visualizations". In: *IEEE transactions on visualization and computer graphics* (2018).
97. Lawrence J Sanna, Norbert Schwarz, and Shevaun L Stocker. "When debiasing backfires: Accessible content and accessibility experiences in debiasing hindsight." In: *Journal of Experimental Psychology: Learning, Memory, and Cognition* 28.3 (2002), p. 497.
98. Ali Sarvghad, Melanie Tory, and Narges Mahyar. "Visualizing dimension coverage to support exploratory analysis". In: *IEEE transactions on visualization and computer graphics* 23.1 (2016), pp. 21–30.
99. Rainer Sauerborn, Peter Berman, and A Nougtara. "Age bias, but no gender bias, in the intra-household resource allocation for health care in rural Burkina Faso". In: *Health transition review* (1996), pp. 131–145.
100. Phoebe Sengers et al. "Reflective design". In: *Proceedings of the 4th decennial conference on Critical computing: between sense and sensibility*. 2005, pp. 49–58.
101. Yedendra Babu Shrinivasan and Jarke J van Wijk. "Supporting the analytical reasoning process in information visualization". In: *Proceedings of the SIGCHI conference on human factors in computing systems*. 2008, pp. 1237–1246.
102. *Slate*. URL: https://slate.org/.
103. Cheryl Staats et al. "State of the Science: Implicit Bias Review 2017". In: *Kirwan Institute for the Study of Race and Ethnicity* (2017).
104. Poorna Talkad Sukumar and Ronald Metoyer. "A Visualization Approach to Addressing Reviewer Bias in Holistic College Admissions". In: *Cognitive Biases in Visualizations*. Springer, 2018, pp. 161–175.
105. James Surowiecki. *The wisdom of crowds*. Anchor, 2005.

106. Poorna Talkad Sukumar, Ronald Metoyer, and Shuai He. "Making a pecan pie: Understanding and supporting the holistic review process in admissions". In: *Proceedings of the ACM on Human-Computer Interaction* 2.CSCW (2018), pp. 1–22.
107. Anthony Tang, Carman Neustaedter, and Saul Greenberg. "Videoarms: embodiments for mixed presence groupware". In: *People and Computers XX—Engage*. Springer, 2007, pp. 85–102.
108. Richard H Thaler and Cass R Sunstein. *Nudge: Improving decisions about health, wealth, and happiness*. Penguin, 2009.
109. Ayanna K. Thomas and Peter R. Millar. "Reducing the framing effect in older and younger adults by encouraging analytic processing". In: *The Journals of Gerontology Series B: Psychological Sciences and Social Sciences* 2.139 (2011).
110. John W Tukey. "We need both exploratory and confirmatory". In: *The American Statistician* 34.1 (1980), pp. 23–25.
111. Amos Tversky and Daniel Kahneman. "Availability: A heuristic for judging frequency and probability". In: *Cognitive Psychology* 5.2 (1973), pp. 207–232.
112. Amos Tversky and Daniel Kahneman. "The framing of decisions and the psychology of choice". In: *Science* 211 (1981), pp. 453–458.
113. Supreme Court of the United States. *Students for Fair Admissions v. Harvard*. 2023. URL: https://www.supremecourt.gov/opinions/22pdf/20-1199_hgdj.pdf.
114. André Calero Valdez, Martina Ziefle, and Michael Sedlmair. "Priming and anchoring effects in visualization". In: *IEEE Transactions on Visualization and Computer Graphics* 24.1 (2018), pp. 584–594.
115. Emily Wall, John Stasko, and Alex Endert. "Toward a design space for mitigating cognitive bias in vis". In: *2019 IEEE Visualization Conference (VIS)*. IEEE. 2019, pp. 111–115.
116. Emily Wall et al. "A formative study of interactive bias metrics in visual analytics using anchoring bias". In: *IFIP Conference on Human-Computer Interaction*. Springer. 2019, pp. 555–575.
117. Emily Wall et al. "A Markov Model of Users' Interactive Behavior in Scatterplots". In: *2019 IEEE Visualization Conference (VIS)*. IEEE. 2019, pp. 81–85.
118. Emily Wall et al. "Four perspectives on human bias in visual analytics". In: *Cognitive biases in visualizations*. Springer, 2018, pp. 29–42.
119. Emily Wall et al. "Left, Right, and Gender: Exploring Interaction Traces to Mitigate Human Biases". In: *arXiv preprint* arXiv:2108.03536 (2021).
120. Emily Wall et al. "Podium: Ranking Data Using Mixed-Initiative Visual Analytics". en. In: *IEEE Transactions on Visualization and Computer Graphics* 24.1 (Jan. 2018), pp. 288–297. ISSN: 1077-2626. DOI: https://doi.org/10.1109/TVCG.2017.2745078. URL: http://ieeexplore.ieee.org/document/8019863/ (visited on 04/09/2023).
121. Emily Wall et al. "Warning, bias may occur: A proposed approach to detecting cognitive bias in interactive visual analytics". In: *2017 IEEE Conference on Visual Analytics Science and Technology (VAST)*. IEEE. 2017, pp. 104–115.
122. Monica L Wang et al. "A Systematic Review of Diversity, Equity, and Inclusion and Antiracism Training Studies: Findings and Future Directions". In: *Translational Behavioral Medicine* 14.3 (Oct. 2023), pp. 156–171. ISSN: 1613-9860. DOI: https://doi.org/10.1093/tbm/ibad061. URL: http://dx.doi.org/10.1093/tbm/ibad061.
123. Qianwen Wang et al. "Visual analysis of discrimination in machine learning". In: *IEEE Transactions on Visualization and Computer Graphics* 27.2 (2020), pp. 1470–1480.
124. Xiaomei Wang et al. "Cognitive Work Analysis and Visualization Design for the Graduate Admission Decision Making Process". In: *Proceedings of the Human Factors and Ergonomics Society Annual Meeting* 64.1 (Dec. 2020), pp. 815–819. ISSN: 2169-5067. DOI: https://doi.org/10.1177/1071181320641189. URL: http://dx.doi.org/10.1177/1071181320641189.

125. Wesley Willett, Jeffrey Heer, and Maneesh Agrawala. "Scented widgets: Improving navigation cues with embedded visualizations". In: *IEEE Transactions on Visualization and Computer Graphics* 13.6 (2007), pp. 1129–1136.
126. *Women in State Legislatures in 2019.* https://www.ncsl.org/legislators-staff/legislators/womens-legislative-network/women-in-state-legislatures-for-2019.aspx Accessed 2021-03-04.
127. Richard Zemel et al. "Learning fair representations". In: *30th International Conference on Machine Learning, ICML 2013* (Jan. 2013), pp. 1362–1370.
128. Brian Hu Zhang, Blake Lemoine, and Margaret Mitchell. "Mitigating unwanted biases with adversarial learning". In: *AAAI/ACM Conference on AI, Ethics, and Society*. 2018, pp. 335–340.

Part IV
Ongoing and Future Work

Next Steps in Research on Human Bias in Visual Data Analysis

8

8.1 Bias Propagation as an Echo Chamber

Without techniques for identifying and guarding against biased input, applications run the risk of showing users biased views of their data that correspond to what they want to use, rather than truthful representations of the information.

For example, in model-steering situations, user input guides analytic models to focus on salient aspects of the domain being studied [15]. Without guarding against potentially biased user input, the system may overfit the model to the biased input. **The result may be a system that shows users the views they want to see, but is essentially an "echo chamber" for their own biases**. In fact, recent work by Glickman and Shalot [18] demonstrates dangerous feedback loops that may form in systems with human–machine teams. This serves as a case in point illustration for the urgency in proactively addressing bias in visual data analysis.

8.2 Multiplicity of Bias Definitions

Research described in Sect. 3.2 provides three perspectives on how the term bias is used in visual data analysis. This work begins to create a structure around the multiplicity in definitions of "bias" that sometimes leads to challenges in sharing and collaboration due to a lack of common ground. When considering all of these perspectives, the space in which to study bias in visual data analysis increases dramatically. However, this work is based on observation, without a rigorous methodology for distilling these perspectives. Particularly with recent efforts to capture and mitigate ML biases, we observe a stark trend in the multitude of ways that people use the term "bias" in computing research. As described in Sect. 3.2, bias can be used to describe cognitive limitations of people [24], perceptual predispositions [25], and culturally embedded tendencies to stereotype [19]. However, we also observe instances in computing where bias is used to refer to deviations from target

E. Wall, *Human Bias in Visual Data Analysis*, Synthesis Lectures on Visualization,
https://doi.org/10.1007/978-3-032-09307-3_8

populations in data collection [23], imbalances in datasets [36], and departures in ML model predictions for data based on protected attributes [41], among others. Furthermore, these uses of the term "bias" are often colloquial in nature, in some cases, lacking operational definition, mathematical formulations, examples, or reference to prior definitions. **We see a critical need to systematically survey the ways that the term bias is used in computing, to consolidate overlapping efforts in a comprehensive framework**.

Consider the following example to illustrate the importance of precisely defining the term "bias." The bias metrics described in Sect. 5.2 and used in subsequent systems and studies in Sects. 7.4–7.5 are formulated based on comparing user interactions to a proportional baseline—that is, it assumes that bias occurs when interactions are not evenly distributed across the data. Given that visualizing these metrics in real time sometimes resulted in changes in behavior and decisions, it begs the question whether this was the *right* way to nudge participants. For instance, in Sect. 7.5, some participants expressed explicit goals to choose representative (or proportional) samples of political committees, some aimed to choose equal samples (e.g., equal numbers of Democrats and Republicans), while still others intentionally biased their selections in other ways (e.g., all female politicians). Each of these could be mathematically formulated as different baseline models of what constitutes unbiased behavior, what has further ethical implications, in that designers must take on the social responsibility to choose visualization designs and bias computation mechanisms that reflect social values without unduly compromising user agency. Precise and rigorous definitions of bias are needed to distinguish these otherwise terminologically similar concepts.

8.3 Is Bias Bad?

In short, **it depends on the perspective on bias that we adopt**. This makes the previous discussion on multiplicity of bias definitions particularly critical.

The word bias itself has a negative connotation. It evokes a sense of imperfection that we tend to think we can overcome with careful critical thinking and reflection. If we consider the ways that bias is often portrayed in the media [1], it is typically a way of framing critical failures, most recently in AI. In these situations, bias is an illusive concept underlying a multitude of nefarious outcomes that ultimately boil down to machines finding a way to capture the worst of humanity. When biased algorithms deny life-saving healthcare [16] or reflect racist societal contempt [28], it would be generous to say that bias is bad. Similarly, when people make unfair decisions driven by religious stereotypes, as in the case of the fingerprint analyst who misidentified Brandon Mayfield as the man responsible for the Madrid train bombings due to confirmation bias after Mayfield converted to the Islam religion [14, 35], it is likewise indisputable that bias is bad. However, these instances are not wholly representative of the concept of bias.

We emphasize that not all bias is bad. From the perspective that bias is a filter and the perspective that bias is a preconception (see Sect. 3.2), it can be both beneficial and detrimental depending on circumstances. Perceptual filters prevent us from experiencing information overload. Bias is what enables us to hear our names mentioned among the noise in a crowded room. It keeps us safe when we perceive a dog with a wagging tail differently than one with a snarl and attacking posture. It empowers us to make quick calculations and associations based on patterns of information we have seen before. However, they can also cause us to inadvertently filter out information relevant to a given decision. Unconscious biases like innate risk-aversion tendencies can help us to make deliberate, mindful decisions, but on the other side of the spectrum can lead to impulsive high-risk decisions. Likewise, the same heuristic approach to problem-solving that produces cognitive biases is what allows us to not be bogged down by constant trivial decisions. It allows us to solve problems more quickly and to make fast perceptual judgments.

This highlights the need for future work on bias to clearly define the term as intended in a given project, through referencing the papers where an adopted use of the term bias originated, formulating a unique mathematical definition, providing intuitive examples, and so on.

8.4 Advancing Models of Bias Identification

A wide array of methods have been recently developed for human bias detection, including application of statistical and machine learning models in visual data analysis [17, 20, 29, 30, 39] (as in Sect. 5.2) and passive measures of visual attention via eyetracking to capture attentional biases [5, 26]. However, each of these models are typically developed in the context of a singular interface, dataset, and task with a small number of users. They typically focus on a relatively narrow set of signals and excel at capturing different notions of bias, related to, e.g., uneven distributions of interactions [39] or lack of diversity in data exploration behaviors [17]. **There is opportunity to build upon these efforts to develop advanced models of bias identification**.

First, the controlled contexts in which these metrics are primarily developed and studied are not reflective of the range of tasks in visual data analysis. In reality, visual data analysis can be much messier with analysts examining alternative hypotheses and switching between potentially very different subtasks in diverse analytic interfaces. Robust models of bias identification in visual data analysis will require a greater understanding of how bias materializes in various interfaces and analytic subtasks (e.g., ranking, clustering, etc.) as well as how these subtasks combine into more complete sensemaking. Similarly, models will need to deal with heterogeneous data types that are common throughout visual data analysis. Whereas most existing attentional measures are built based on viewing images or videos [5, 26] and most discrete bias measures are built on tabular numerical and categorical data [17, 39], there are other sources of data, namely text, that are often the object of visual data analysis [9, 10].

Another potential improvement to the metrics we observe would be to develop a temporal weighting scheme, where all discrete and passive interactions are used to compute the bias metrics, but the interactions are weighted as a function of recency. The most recent interactions would be weighted more highly than those performed early in the user's analysis process, to allow metrics to better capture instances of user correction. Finally, there is opportunity to explore composite models that incorporate a range of existing metrics to improve predictive performance. Hence models may be better equipped to accurately gauge instances of bias by including signals of both discrete interaction behaviors (clicks, hovers, scrolls, etc.) alongside passive measures of attention via eye tracking.

8.5 Expertise, Preference, and Contextual Factors Can Confound Bias

Visual analytic applications continue to model users and adapt interfaces, visualizations, and analytic models based on their interactions. How can these systems differentiate between valuable subject matter expertise (which should be incorporated), and biased input? In the analytic process, humans have intuition and expertise to guide them. However, **the interaction patterns of expert analysts and cognitively biased analysts might look very similar despite very different cognitive processes**. Consider the case of an analyst focusing his attention on evidence surrounding a particular suspect. Such focus may result from cognitive bias, or it may result from quick deliberate decisions based on years of experience. The analyst might also have knowledge about the case not captured by the data at the time, like breaking new evidence. Thus, it is important to understand the role context and domain expertise play in structuring the visual analytic process to differentiate expertise from cognitive biases producing an inferior analytic process.

One approach for making the distinction between valuable domain expertise and biased input might be to consider the consistency or inconsistency of a user's interaction sequences. More sophisticated approaches could be derived by studying the differences in interaction sequences of domain experts and novices who are biased. It may also be useful to study large groups of users, expert or novice, modeling their processes and biases, to provide additional context to the machine intelligence about ranges of typical and outlier behaviors.

User annotations of their own interactions would be another possibility for improving the machine's ability to distinguish expert and biased behavior. This would facilitate creating a common understanding between the system and user by eliciting explicit user feedback and reflection. Metrics for capturing bias could then be adjusted in real time to weigh subsequent interactions accordingly, so that confounding factors are not confused as negative biases. However, this approach would only work well for cases where users have heightened metacognitive awareness—otherwise, users will be susceptible to a bias blind spot, where they are unable to recognize instances of their own biases [32].

Another confound can be the distinction between conscious and unconscious biases, as we observed in the study on political decision-making for abortion bans [38] in Sect. 7.5. We observed some individuals made nuanced choices, explicitly biased and balanced in different ways. For instance, some individuals chose balanced committee compositions based on equal numbers of representatives by gender and political affiliation, but nonetheless intentionally chose individuals whose position on abortion aligned with their own, giving a false sense of balance to the committee. Like expertise, preference, and contextual factors, the interactions users perform related to conscious or unconscious bias may look similar from a behavioral perspective, yet it may render interventions wholly ineffective.

8.6 False Positives Versus False Negatives

Given the imperfection of quantifying bias from user interactions and the numerous potential confounds as described above, it begs the question: what is the harm of inaccuracy? A false positive (i.e., the system believes you are biased when you are not) could be frustrating to users, but we posit may be otherwise relatively harmless. For example, users might intentionally choose to not interact with an attribute because it is either not important or they are not interested in it (e.g., GPA in college admissions decisions). Labeling these as underemphasized or biased may not be correct, as it was a conscious decision by users to ignore them. On the other hand, a false negative (i.e., the system believes you are not biased when you are) could be much more harmful, leading to unchecked errors. Furthermore, "false negative" circumstances are essentially the present norm, given that most systems do not attempt to capture bias in the analysis process. In such circumstances, biases would have propagated unchecked regardless. In cases where decisions carry heavy importance, such as college admissions, bail in the criminal justice system, or intelligence analysis, **false positives may be the preferred direction to err**.

8.7 Personalizing Intervention

The visualization community is collectively producing a wealth of knowledge about individual differences in the ways people interact with data [27]. In particular, research has revealed differences in observable interaction paths [4, 7, 13] and gaze patterns [7] for individuals with different personality traits, reasoning strategies, and susceptibility to biases. For instance, in Sect. 5.6, we described how individuals with high versus low values of personality traits such as conscientiousness, extraversion, agreeableness, and neuroticism displayed different trends in their interaction strategies and also displayed weak correlations with their susceptibility to Dunning–Kruger effect. These differences highlight an opportunity for future work to develop personalized interventions.

Personalized interventions may leverage individual differences like personality traits to introduce interventions to promote effective decision-making, in ways that individuals are uniquely more inclined to respond to. For instance, individuals high on the scale of "neuroticism" could be more receptive and responsive to interventions that simply provide information (increase awareness) about potential biases in their process and will self-correct accordingly, whereas individuals low on the scale of "neuroticism" may prefer more intrusive [6, 37] interventions instead, where the individuals do not need to tend to bias information as closely. To this end, combining work on individual differences [27] alongside literature on guidance [6] in visualization and reflective design [34] in HCI may be fruitful starting points for designing future visualization systems that may help users become more capable of accurate self-assessment and self-correction.

8.8 Metacognition and Behavior Change: New Frontiers for Intervention

The shift toward automated and semi-automated systems in the age of AI has brought with it a slew of AI failures and instances of harmful biases, and accordingly the value proposition for human input has become evident. Thus far, methods for mitigating biases have relied on low-tech training courses to educate analysts about cognitive pitfalls [22] or addressed the propagation of these biases through automated techniques [2, 40].

To future-proof against AI failures and breakdowns resulting from human biases requires more than what out-of-context training courses or even automated mitigation strategies can offer. What we need is a world where analysts and decision-makers are equipped to assess their own limitations and act accordingly. The vast majority of educational content focuses on developing technical skillsets, neglecting the importance of training strong metacognitive awareness and strategies [21].

An ideal future is one where biased behaviors and decisions are caught before they are injected into systems and propagated downstream. To do so, one promising direction is to develop metacognitive interventions—both in educational content, as a set of transferable skills which can be taught [33] and as in-the-moment interventions that can reinforce metacognitive awareness and strategy [8]. In particular, these interventions can trigger metacognitive reflection, or reflection about one's own thought process, empowering analysts to self-correct.

However, while awareness of and reflection on biases may be sufficient for many to address biased analysis patterns, it may be insufficient for others. Thus, recent work on behavior change may be another promising framework for bias mitigation [11, 12]. In this work, theories from cognitive and clinical psychology are used to help people improve behaviors, e.g., related to smoking cessation [3, 31]. Translated into the context of visual data analysis, these theories offer further promise toward empowering responsible data analysis practices.

References

1. *AI Incident Database*. URL: https://incidentdatabase.ai.
2. Rachel Bellamy et al. "AI Fairness 360: An Extensible Toolkit for Detecting and Mitigating Algorithmic Bias". In: *IBM Journal of Research and Development* PP (Sept. 2019), pp. 1–1. DOI: https://doi.org/10.1147/JRD.2019.2942287.
3. Belinda Borrelli and Robin Mermelstein. "Goal setting and behavior change in a smoking cessation program". In: *Cognitive Therapy and Research* 18.1 (1994), pp. 69–83.
4. Eli T Brown et al. "Finding waldo: Learning about users from their interactions". In: *IEEE Transactions on visualization and computer graphics* 20.12 (2014), pp. 1663–1672. DOI: https://doi.org/10.1109/TVCG.2014.2346575.
5. Benjamin T Carter and Steven G Luke. "Best practices in eye tracking research". In: *International Journal of Psychophysiology* 155 (2020), pp. 49–62.
6. Davide Ceneda et al. "Characterizing guidance in visual analytics". In: *IEEE Transactions on Visualization and Computer Graphics* 23.1 (2017), pp. 111–120.
7. Mengyu Chen and Emily Wall. "Perception of Skill in Visual Problem Solving: An Analysis of Interactive Behaviors, Personality Traits, and the Dunning-Kruger Effect". In: *TREX: Workshop on TRust and EXpertise in Visualization* (2022).
8. Mengyu Chen et al. "A Novel Lens on Metacognition in Visualization". In: (2025).
9. Yanan Da et al. "Visual Salience to Mitigate Gender Bias in Recommendation Letters". In: *Workshop on Visualization for Social Good (VIS4Good, at VIS'23)* (2023).
10. Yanan Da et al. "Visual Salience to Mitigate Gender Bias in Recommendation Letters". In: *IFIP Conference on Human-Computer Interaction*. Springer. 2025.
11. Ziwei Dong et al. "A Design Space of Behavior Change Interventions for Responsible Data Science". In: *Proceedings of the 30th International Conference on Intelligent User Interfaces*. IUI '25. Association for Computing Machinery, 2025, pp. 37–53. ISBN: 9798400713064. DOI: https://doi.org/10.1145/3708359.3712140. URL: https://doi.org/10.1145/3708359.3712140.
12. Ziwei Dong et al. "Behavior Matters: An Alternative Perspective on Promoting Responsible Data Science". In: *Proceedings of the ACM Conference on Computer Supported Cooperative Work & Social Computing* (2025).
13. Wenwen Dou et al. "Recovering Reasoning Process From User Interactions". In: *IEEE Computer Graphics & Applications* May/June.March (2009), pp. 52–61. URL: http://citeseerx.ist.psu.edu/viewdoc/download?doi=10.1.1.157.407%7B%5C&%7Drep=rep1%7B%5C&%7Dtype=pdf.
14. Itiel Dror. "Combating bias: The next step in fighting cognitive and psychological contamination". In: *Journal of Forensic Sciences* 57.1 (2012), pp. 276–277.
15. A Endert et al. "The State of the Art in Integrating Machine Learning into Visual Analytics". In: *Computer Graphics Forum*. Wiley Online Library. 2017.
16. Virginia Eubanks. *Automating inequality: How high-tech tools profile, police, and punish the poor*. St. Martin's Press, 2018.
17. Mi Feng, Evan Peck, and Lane Harrison. "Patterns and Pace: Quantifying Diverse Exploration Behavior with Visualizations on the Web". In: *IEEE transactions on visualization and computer graphics* 25.1 (2019), pp. 501–511.
18. Moshe Glickman and Tali Sharot. "Biased AI systems produce biased humans". In: (2022).
19. Anthony G Greenwald and Linda Hamilton Krieger. "Implicit bias: Scientific foundations". In: *California Law Review* 94.4 (2006), pp. 945–967.
20. Sunwoo Ha et al. "A unified comparison of user modeling techniques for predicting data interaction and detecting exploration bias". In: *IEEE Transactions on Visualization and Computer Graphics* 29.1 (2022), pp. 483–492.

21. Diane F Halpern. "Teaching critical thinking for transfer across domains: Disposition, skills, structure training, and metacognitive monitoring." In: *American psychologist* 53.4 (1998), p. 449.
22. Richards J. Heuer Jr. *Psychology of Intelligence Analysis*. Washington, DC, 1999.
23. Justin Jager, Diane L Putnick, and Marc H Bornstein. "II. More than just convenient: The scientific merits of homogeneous convenience samples". In: *Monographs of the Society for Research in Child Development* 82.2 (2017), pp. 13–30.
24. Daniel Kahneman. *Thinking, fast and slow*. Macmillan, 2011.
25. Kurt Koffka. *Principles of Gestalt Psychology*. Vol. 44. Routledge, 2013.
26. Jukka M Leppänen. "Using eye tracking to understand infants' attentional bias for faces". In: *Child development perspectives* 10.3 (2016), pp. 161–165.
27. Zhengliang Liu, R Jordan Crouser, and Alvitta Ottley. "Survey on individual differences in visualization". In: *Computer Graphics Forum*. Vol. 39. 3. Wiley Online Library. 2020, pp. 693–712.
28. Josh Lovejoy and Jess Holbrook. *Human-Centered Machine Learning*. URL: https://medium.com/google-design/human-centered-machine-learning-a770d10562cd.
29. Shayan Monadjemi, Roman Garnett, and Alvitta Ottley. "Competing models: Inferring exploration patterns and information relevance via bayesian model selection". In: *IEEE Transactions on Visualization and Computer Graphics* 27.2 (2020), pp. 412–421.
30. Alvitta Ottley, Roman Garnett, and Ran Wan. "Follow the clicks: Learning and anticipating mouse interactions during exploratory data analysis". In: *Computer Graphics Forum*. Vol. 38. 3. Wiley Online Library. 2019, pp. 41–52.
31. Jeni Paay et al. "Understanding Individual Differences for Tailored Smoking Cessation Apps". In: *Proceedings of the 33rd Annual ACM Conference on Human Factors in Computing Systems*. CHI '15. Seoul, Republic of Korea: Association for Computing Machinery, 2015, pp. 1699–1708. ISBN: 9781450331456. DOI: https://doi.org/10.1145/2702123.2702321. URL: https://doi.org/10.1145/2702123.2702321.
32. Emily Pronin, Daniel Y Lin, and Lee Ross. "The bias blind spot: Perceptions of bias in self versus others". In: *Personality and Social Psychology Bulletin* 28.3 (2002), pp. 369–381.
33. Gregory Schraw. "Promoting general metacognitive awareness". In: *Instructional science* 26.1 (1998), pp. 113–125.
34. Phoebe Sengers et al. "Reflective design". In: *Proceedings of the 4th decennial conference on Critical computing: between sense and sensibility*. 2005, pp. 49–58.
35. Robert B Stacey. "A Report on the Erroneous Fingerprint Individualization in the Madrid Train Bombing Case". In: *Journal of Forensic Identification* 54.6 (2004), pp. 706–718.
36. Antonio Torralba and Alexei A Efros. "Unbiased look at dataset bias". In: *CVPR 2011*. IEEE. 2011, pp. 1521–1528.
37. Emily Wall, John Stasko, and Alex Endert. "Toward a design space for mitigating cognitive bias in vis". In: *2019 IEEE Visualization Conference (VIS)*. IEEE. 2019, pp. 111–115.
38. Emily Wall et al. "Left, Right, and Gender: Exploring Interaction Traces to Mitigate Human Biases". In: *arXiv preprint* arXiv:2108.03536 (2021).
39. Emily Wall et al. "Warning, bias may occur: A proposed approach to detecting cognitive bias in interactive visual analytics". In: *2017 IEEE Conference on Visual Analytics Science and Technology (VAST)*. IEEE. 2017, pp. 104–115.
40. Brian Hu Zhang, Blake Lemoine, and Margaret Mitchell. "Mitigating unwanted biases with adversarial learning". In: *AAAI/ACM Conference on AI, Ethics, and Society*. 2018, pp. 335–340.
41. Zhe Zhang and Daniel B Neill. "Identifying significant predictive bias in classifiers". In: *arXiv preprint* arXiv:1611.08292 (2016).

MIX
Papier aus verantwortungsvollen Quellen
Paper from responsible sources
FSC® C105338

If you have any concerns about our products,
you can contact us on
ProductSafety@springernature.com

In case Publisher is established outside the EU,
the EU authorized representative is:
Springer Nature Customer Service Center GmbH
Europaplatz 3, 69115 Heidelberg, Germany

Printed by Libri Plureos GmbH
in Hamburg, Germany